"Yana Manyukhina's beautifully clear book will be of value to all concerned with the politics of food and consumerism. It analyses not only how people develop as ethical food consumers but also, perhaps more crucially, why they make the life-style changes that are so urgently needed to promote sustainable ways of living."

Priscilla Alderson, Professor Emerita,
University College London

Ethical Consumption: Practices and Identities

This book engages with the topic of ethical consumption and applies a critical-realist approach to explore the process of becoming and being an ethical consumer. By integrating Margaret Archer's theory of identity formation and Christian Coff's work on food ethics, it develops a theoretical account explicating the generative mechanism that gives rise to ethical consumer practices and identities. The second part of the book presents the findings from a qualitative study with self-perceived ethical food consumers to demonstrate the fit between the proposed theoretical mechanism and the actual experiences of ethically committed consumers. Through integrating agency-focused and socio-centric perspectives on consumer behaviour, the book develops a more comprehensive and balanced approach to conceptualising and studying consumption processes and phenomena.

Yana Manyukhina gained her PhD in Sociology and Social Policy in 2016 from the University of Leeds, UK.

Routledge Studies in Critical Realism

Critical Realism is a broad movement within philosophy and social science. It is a movement that began in British philosophy and sociology following the founding work of Roy Bhaskar, Margaret Archer and others. Critical Realism emerged from the desire to realise an adequate realist philosophy of science, social science, and of critique. Against empiricism, positivism and various idealisms (interpretivism, radical social construction-ism), Critical Realism argues for the necessity of ontology. The pursuit of ontology is the attempt to understand and say something about 'the things themselves' and not simply about our beliefs, experiences, or our current knowledge and understanding of those things. Critical Realism also argues against the implicit ontology of the empiricists and idealists of events and regularities, reducing reality to thought, language, belief, custom, or experience. Instead Critical Realism advocates a structural realist and causal powers approach to natural and social ontology, with a focus upon social relations and process of social transformation.

Important movements within Critical Realism include the morphogenetic approach developed by Margaret Archer; Critical Realist economics developed by Tony Lawson; as well as dialectical Critical Realism (embracing being, becoming and absence) and the philosophy of metaReality (emphasising priority of the non-dual) developed by Roy Bhaskar.

For over thirty years, Routledge has been closely associated with Critical Realism and, in particular, the work of Roy Bhaskar, publishing well over fifty works in, or informed by, Critical Realism (in series including Critical Realism: Interventions; Ontological Explorations; New Studies in Critical Realism and Education). These have all now been brought together under one series dedicated to Critical Realism.

The Centre for Critical Realism is the advisory editorial board for the series. If you would like to know more about the Centre for Critical Realism, or to submit a book proposal, please visit www.centreforcriticalrealism.com.

Sociology, Health and the Fractured Society
A Critical Realist Account
Graham Scambler

Empiricism and the Metatheory of the Social Sciences
Roy Bhaskar

Ethical Consumption: Practices and Identities
A Realist Approach
Yana Manyukhina

For more information about this series, please visit:
https://www.routledge.com/Routledge-Studies-in-Critical-Realism-Routledge-Critical-Realism/book-series/SE0518

Ethical Consumption: Practices and Identities

A Realist Approach

Yana Manyukhina

LONDON AND NEW YORK

First published 2018 by Routledge

2 Park Square, Milton Park, Abingdon, Oxfordshire OX14 4RN
52 Vanderbilt Avenue, New York, NY 10017

Routledge is an imprint of the Taylor & Francis Group, an informa business

First issued in paperback 2019

British Library Cataloguing-in-Publication Data
A catalogue record for this book is available from the British Library

Library of Congress Cataloging-in-Publication Data
Names: Manyukhina, Yana, author.
Title: Ethical consumption: practices and identities: a realist approach /
Yana Manyukhina.
Description: 1 Edition. | New York: Routledge, 2018. | Series: Routledge
studies in critical realism | Includes bibliographical references and index.
Identifiers: LCCN 2018002595 | ISBN 9781138895539 (hbk) |
ISBN 9781315179476 (ebk)
Subjects: LCSH: Consumption (Economics)–Moral and ethical aspects. |
Consumption (Economics)–Social aspects. | Consumer behavior–Moral and
ethical aspects. | Consumers–Attitudes.
Classification: LCC HB835 .M336 2018 | DDC 174/.4–dc23
LC record available at https://lccn.loc.gov/2018002595

ISBN: 978-1-138-89553-9 (hbk)
ISBN: 978-0-367-82110-4 (pbk)

Typeset in Times New Roman
by Deanta Global Publishing Services, Chennai, India

I dedicate this book to all those who believe in a better world and are not afraid to pursue it.

Contents

Foreword

Margaret S. Archer

This is a wonderful first book, theoretically robust and empirically sensitive. Without stating it explicitly, from start to finish, Yana Manyukhina provides a comprehensive rebuttal of one of Bourdieu's central, but least quoted assumptions, namely the first sentence of *The Logic of Practice*: "Of all the oppositions that artificially divide social science, the most fundamental, and the most ruinous, is the one that is set up between subjectivism and objectivism" (1990 [1980] p. 25).

Instead, people have "concerns", things that matter to them, which reflexively they devote time to thinking over in their internal conversations and devising courses of action considered (fallibly) as likely to attain them. This is partly what makes agents human; equally, such concerns differ and thus contribute to making one human being different from another – usually called their personal identities. This statement has to be complemented by another. At any given time, people are situated somewhere, however mobile they may be and however far they have moved (or been moved), from their natal origins. In turn, where they are placed – geographically, culturally, and socially – affects the actions they can entertain and accomplish because there is no such thing as context-less action. In sociology, this is commonly abbreviated into "the Structure-Agency problem", which was popular 50 years ago and some claimed that they had "transcended" it.

Such claimants (Giddens and Bourdieu being the best known) were simply guilty of "central conflation", by maintaining that the two elements were mutually constitutive and (hence, though it does not follow) were also ontologically inseparable. Others gave allegiance to one side or the other. Residual holists (often neo-Marxists) continued the "downwards conflationist" tradition, according causal hegemony to structural or cultural factors, whilst "upwards conflation" was the resort of the remaining protagonists of Methodological Individualism (most influential in the guise of Rational Choice Theory). Only Critical Realists upheld the analytical separability of Structure and Agency, despite their interplay and interdependence, thus setting a new research agenda focussing upon *how* they were interrelated and what *mediated* between them because vague talk about their mutual conditioning failed to supply any generative mechanism.

Confronting this central problem, and the ensuing theoretical morass, is one of the major strengths of Dr Manyukhina's book, which she sustains from beginning to end. Particularly engrossing is the activity upon which she focuses, "ethical consumption" (largely ethical provisioning and eating). It is quotidian yet demanding to commit to living as a vegan, vegetarian, or even a consistent consumer of organic produce alone. It is neither her own dietary regime, nor does it have the glamour of those whose prime concern is proclaimed as "peace", "justice" or "equality", all of which are diffuse in their demands. What is impressive in the first chapters is her illustration of how the "Structure–Agency debate" has impinged on the surprisingly large corpus of literature now surrounding "ethical eating". First, she presents a general and economic critique of the three forms of conflation in this context and then proceeds to a diagnosis of their particular shortcomings in relation to "ethical eaters" .

Second, she never forgets or downplays the fact that we all live in the real world, constituted by three orders: nature, material culture, and the social. The latter is important but can never subsume the other two orders of natural reality. In this, her work parallels Colin Campbell's *Myth of Social Action*, where he shows all action being gradually assimilated by a succession of authors to the social domain. Instead, nature, as it were, is shown rudely to rebuke certain tyro enthusiasts in ethical eating about their indifference to their health. It constitutes a challenge to their reflexivity and does not determine the failure of their concern. Rather, it "requires" reflexive redefinition of their initial project in relation to their concern. Since Yana is dealing with singular subjects (though I see no reason why later research could not be extended to collective subjects, such as families, schools, or even alternative governmental "healthy eating" projects) their responses are not identical: some give up for a time, others turn to becoming better informed about nutrition, and still others to compromise projects.

What Yana Manyukhina wants above all is to elaborate a "concerns-based" approach, grounded upon what matters most to her singular subjects, making for their diversity: how it originated, what channelled this caring into a particular form, what challenged them and sometimes derailed them, why some picked themselves up and sometimes modified their dietary projects whilst others renewed their prior dedication. All of this is held to depend upon the subject's reflexivity, and I find her argument compelling, especially her detailed treatment of the strengthening or mutation of concerns if compared with the mysterious origins and workings of Humean "preferences".

Structure and culture contextualise this everyday decision-making at every turn and since constraints require something to constrain and enablements something to enable, she consistently shows that what they impinge upon are subjects' projects and the conditions required for turning them into established practices, constitutive of a distinctive *modus vivendi*. Some of these constraints are well-known, such as the higher cost of "ethical goods", estimated at 45 per cent, and obviously constrain food consumption for students and those on lower incomes. Yet, even here, Yana shows how constraints do not bludgeon her subjects into a uniform response. Despite her small investigative group of 10, it is more than sufficient to demonstrate various reflexive strategies of evasion. For example, some grow their own produce, some form or join co-operatives for bulk buying and others scour their vegetarian recipe books for substitute products or alternative dishes.

Yana's theoretical point is that this variety of responses remains inexplicable if a constraint is treated as a blunt instrument, affecting all in the same way, regardless of their subjectivity. Empirically, her practice of accompanying her interviewees on their shopping trips and listening to their running commentaries enriched her account of their tactics of circumvention. Certainly, some social scientists would reply correctly that opportunity costs were still entailed, but this would hinge on trade-offs between money and time and the use of multiple currencies (for time-banks exist), which merely opens up the scope for further variations in subjective responses.

Given these personal and even idiosyncratic varieties of response to the same stimuli, it becomes impossible to exclude subjectivity and to furnish purely objective accounts of action that operate in hydraulic terms. Such attempted explanations can describe how social properties and powers (and natural and technological ones) impinge upon agents, but not how they are received by them: their activation, evasion, or suspension. Their subjective filtration through agential concerns is needed to explain what people really do. Without this mediating process that allows for personal powers, Sociology settles for "what most people do most of the time", and that is a retreat into Humean "constant conjunctions". Since subjectivity is indispensable, the common strategy is for such properties *to be imputed by the investigator*, be it the uniform promotion of vested interests, the universal deployment of Instrumental Rationality or the united conformity to *habitus*. In all of these ways, society grows more influential whilst the agent is denuded of personal powers with the exception of their plasticity.

Culturally, the author makes parallel points about "normativity", never presented as a big wave flattening all into social conformity. Subjects were unanimous in their awareness of obvious sources of painful normative clashes and many overtly stated their unwillingness to enter into a partnership with an omnivore, clearly foreseeing the weariness of daily struggles over incompatible values. However, such straightforward avoidance is not possible where pre-established family relations were concerned, especially as becoming an "ethical eater" often began whilst subjects were young and living at home. As someone once said, "commensality is more intimate than connubiality", which seems particularly apposite for interchanges at the family dinner table.

Perhaps I am going too far in extrapolating from one case where the omnivorous father thrust forkfuls of meat in his daughter's face. This seems a case of overt normative conflict that would likely result in more general relational harms being generated between father and daughter. Much more difficult to handle were instances of warm family relationships where the parent(s) provisioning and cooking for some family occasion had made thoughtful propitiatory efforts to offend no-one and to preserve their relational goods by serving free-range chicken and cheesecake. Although this was a bridge too far for her aspirant "ethically eating" daughter, this young person conformed rather than disrupt the family celebration. Yet, note, she was not capitulating to her parents' normative conspectus, but acknowledging their reflexive kindness in trying to satisfy everyone. This example raises a bigger problem about the hierarchy of our concerns.

Rarely do we have one ultimate concern, below which others are subordinated if not eliminated. The whole text shows both how far these subjects would go to protect this particular concern – usually as unobtrusively as possible – but it could not preclude their caring about other matters and there is no formula for guaranteeing complementarity between these concerns. The interviewees are like most of us, slipping, compromising, and living guiltily aware that we are not serving any of our concerns as well as they deserve. A particular difficulty attaching to "ethical eating" is that it is, in my view, too restricted a concern to form the basis of an *entire modus vivendi*. Even those who extend what matters to them upwards to become involved in climate change, still have to work and, ideally, find their employment satisfying and sustainable. It can be done: work for Green Peace, be politically active in environmental causes and animal rights, have a partner of the same persuasions etc., but this combination is not structurally or culturally available to all, let alone throughout their whole lifetimes.

As Yana puts it, "ethical choices and activities are not mere instruments for the construction and presentation of self, but are extensions and expressions of it – not a means to some further end, but are ends in themselves" (p. 179). This is similar to what Harry Frankfurt means by "wholeheartedness". Few succeed in this, but even our failings are sources for reflexive examination, remedial determination, and renewed dedication. As he also maintains, without undertaking such reflexive processes, what we ultimately betray is ourselves.

Preface

Morality is inherent in humans. We all have a moral sense, engage in moral reasoning, pass moral judgments, and we all, if to a different extent and with different levels of success and persistence, construct our lives according to what we believe is the right way to be and to act. Which particular domain becomes the locus of our righteous endeavours depends on what we, as persons, consider to be of the utmost importance, be it in the context of our lives as individuals or as members of the larger community called "humanity". My book seeks to analyse ethical consumption as one increasingly relevant form of moral behaviour, inspired by concerns over one's impact on the many components of our inescapably interdependent, interconnected, and intersubjective reality. Ultimately, my aim is to shed light on the chain of causal processes and interactions leading to consumer moral conversion, and to encourage consideration of the kinds of factors that are responsible for shaping one's becoming and being an ethical consumer.

This work first took shape as a PhD dissertation at the University of Leeds. I am grateful to my research supervisors, Nick Emmel and Lucie Middlemiss, for their expert advice and skilful guidance throughout my doctoral studies. Nick helped me to find my philosophical ground by constantly stretching the limits of my ontological vision and challenging me to reach beyond the observable, the empirical, and the concrete. Lucie, however, helped me to navigate the realm of abstract ideas without losing sight of the unique stories that fill my research, bringing it to life through personal voices, and rendering it important and meaningful for anyone who has ever been concerned with understanding why people behave the way they do.

Before this could happen, however, I had to be persuaded that my sincere attempts at understanding the social world could result in a work worth presenting to the academic community. For this, I shall always be grateful to Farida Vis, my Master's thesis supervisor and mentor, my first co-researcher and co-author, and simply my friend. Farida, I keep you in mind every step of the way, and I am hopeful that you will feel proud of what I have achieved – this would never have happened without you.

I thank Margaret Archer, who took interest in my thesis, treated it to a rigorous, thoughtful, and fair review, and saw its potential to become a book. Our ensuing communication, virtual and face-to-face, has itself been an education; I have no doubt that it will have a long-lasting impact on my formation as a social theorist and researcher.

My greatest debt is to Priscilla Alderson, who welcomed me warmly into UCL Institute of Education's growing community of academics inspired by critical realism and eager to unlock its potential to enable researchers to explore, understand, and promote social change. I thank Priscilla wholeheartedly for setting aside time to comment on an earlier draft of my book and to discuss specific issues with me in person. I admire her selfless efforts to help social scientists understand and apply critical realism, and I thank her for being an inexhaustible source of encouragement, for me and I am sure many others.

I owe more than I can say to my mother, Aida Makhammadova, and my partner, Farid Gasanov, both of whom allowed me to pursue my passion for knowledge by shielding me from so many of life's struggles and hardships.

Finally, I am forever indebted to my research participants for letting me into their lives, for generously sharing their stories, histories, and experiences, and for sincere human relationship on which my research project has thrived.

Introduction

This book is about ethical consumption. Rooted in the counter-culture of the 1960s, the phenomenon has long outgrown associations, often pejorative in nature, with "tree-hugging hippies", radical activism, and necessarily anti-establishment lifestyles, and is no longer confined to the fringes of consumer society. Indeed, the past several decades have witnessed a striking increase in ethical consumption across the world. Such remarkable growth has no doubt been enabled by the progressive expansion of ethical goods into the mainstream markets, whilst itself providing a further incentive for the adoption of the ethical ethos by conventional producers and retailers. The manifestations of the trend are not confined to the purchasing of products with ethical credentials by individual consumers with an elevated sense of moral responsibility. The social salience of ethical consumption is revealed in the proliferation of consumer communities – groups and cooperatives – giving rise to shared meanings, cultures, and practices, and an ever-expanding network of activists and grassroots campaigners advocating the use of consumer power as a means of enacting societal change.

The growth of individual and collective consumer mobilisation around ethical causes has brought increasing public, policy, and academic attention to ethical consumption and various aspects of the phenomenon continue to stimulate discussion and debate among experts in different fields. A key question that firmly holds its place on the agenda of sociological, psychological, economic, and marketing research concerns the motivations behind an individual's adoption of an ethical lifestyle. Reflecting a more general tendency to emphasise the symbolic and expressive aspects of consumption – a legacy of "the cultural turn" swaying over the human and social sciences during the final decades of the 20th century – explanations in terms of identity creation and communication have gained currency in the literature on the topic. Numerous theoretical and empirical accounts have tied ethical consumer choices to identity aspirations and goals, providing rich and nuanced commentary on how individuals construct and actualise desired self-concepts, signal higher social, cultural, and ethical standing, and embark on projects of moral selving through what is broadly perceived as a more responsible consumption behaviour. While claims about the important role of ethical consumption in defining and communicating the self proliferate, little remains known about how ethical consumer identities emerge, evolve, and materialise in behaviour.

Uniquely, this book provides an explanatory account of this process. In it, I take the readers on an investigative journey through the lives of nine individuals making their way towards the ethical consumer identity. I observe them in the privacy of their homes, during family dinners, and on social occasions; I follow them around as they visit allotments, farmer's markets, and quirky food shops; I scrutinise their grocery lists, shopping baskets, and eating routines; I explore their thoughts, concerns, and emotions; I watch them retrace their steps through life and witness how their moral commitments spring into being, flourish, or wither away as their priorities, circumstances, and contexts change over time. I interpret the shifting scenes of the subjects' inner and outer lives through a theoretical lens set to throw light on the inner psychological process individuals go through as they develop the ethical consumer identity. This process, I argue, is one of a continuous reflexive exploration of the self and its relationship to the world by inherently normative and evaluative agents who shape their lives around what truly matters to them and in doing so become the particular individuals they feel they should be.

Adopting a critical realist approach, this book goes beyond simply describing how people create and express their desired identities through ethical consumption. Consistent with the fundamental aim of critical realism, I engage with the underlying causal relations between consumers' commitments, practices, and identities and arrive at a mechanism-based explanation of how individual ethical consumer identities emerge, evolve, and transpire, as well as why they may prosper or shrivel in different contexts and at different times in the course of people's lives. Central to my proposed explanatory account are ideas and concepts brought together in a critical realist theory set forward by Archer (2007), namely, ultimate concerns, internal conversation, and reflexivity. These concepts will come to the foreground in Chapter 3, where they will lead the way towards a theoretical explanation of the emotional and mental processes by which individual ethical consumer identities are produced, but they will keep reverberating throughout the narrative, providing critical reference points for the empirical examination of individual commitments to ethical eating presented subsequently in the book.

While I bring forth an individual-level analysis of the ethical consumer identity, my enquiry extends beyond the private psychological workings of a morally concerned subject. Grounded in critical realism, this book explores the role of both human agency and social structure in the shaping of consumers' ethical self. It situates the inherently subjective process of identity formation within the contexts of objective reality and brings together the multiple individual and systemic powers in which ethical consumer practices and identities are contained. The integration of the agency-focused and socio-centric perspectives on consumer behaviour, inspired and enabled by the critical realist framework, is a much-needed shift in conceptualising and studying consumption phenomena and is a distinctive feature of this book.

Ethical consumption, the identity-building consumer, and critical realism

To date, a distinctive body of knowledge has been amassed about the multi-layered relationship between the consumer and the consumed, on both individual and social levels. Since the 1980s, persistent enquiries into the subjective meanings surrounding consumers' acquisition, use, and disposal of commodities have been drawing attention to the symbolic, emotional, and normative charge of consumption, from intentionally conspicuous to the most ordinary and mundane, consistently emphasising the agency, identity, and self-expression involved in consumer acts and activities (Belk, 1988; Miller, 1998; Woodward, 2003, 2007, among many others). Consumer Culture Theory has been a continuous inspiration for this exponentially growing body of work and played a critical role in taking the analysis of consumer identities to the social level. A rich vein of research has emerged linking the ideas and actions of individual consumers to different levels of structure (Dobscha & Ozanne, 2001; Oswald, 1999; Wallendorf, 2001), raising important questions concerning the forging of shared beliefs, meanings, practices, and collective consumer identifications ensuing from them (Holt, 1997; Kozinets, 2002; Muniz & O'Guinn, 2001), and exploring the role of the marketplace as a source of symbolic materials for the production of individual and communal consumer identities (Askegaard & Kjeldgaard, 2002; Coskuner-Balli & Thompson, 2009; Coulter, Price, & Feick, 2003; Gopaldas, 2014; Holt, 2002).

It is beyond the scope of this book to provide a detailed review of the many valuable findings that have emanated from this body of research. I refer to it only to emphasise the remarkable interest and attention that the questions about the role of consumption in identity formation and expression have attracted over the past several decades, and the amount of intellectual effort that has gone into their unravelling. The identity has indeed become the "Rome to which all discussions of modern Western consumption lead" (Gabriel & Lang, 2006, p. 79). It is thus hardly surprising that identity-centred readings of consumer behaviour have spilled over into the subject area of ethical consumption, steering the focus of theoretical and empirical concern towards the figure of an expressive, meaning-seeking, and self-creating consumer. The representation of ethical consumption as a means of creating and expressing individual and social identities presents a lot of opportunities for important sociological explorations. Yet, I argue that if we are truly to interrogate the identity meanings underlying ethical consumer behaviours, we need to understand how the shaping of identity within and around ethical consumption takes place, that is how people develop a sense of self as morally responsible consumers, how these self-concepts materialise in consumption, and how, once achieved, the desired identities are sustained over time.

This book illuminates these questions through a theoretical and empirical analysis of ethical food consumption. Guided by the concept of identity as a unique constellation of a person's ultimate concerns about the world (Archer, 2007), I explore what brings people to care about the ethics of eating, and how these concerns translate into moral commitments that not only direct individuals' actions up to the most mundane choices, such as where to purchase groceries and which brand of coffee to go for, but also profoundly affect their self-image. The grounding of my account in the idea of identity as a unique constellation of a person's most cherished concerns is neither inconsequential nor arbitrary. It is predicated on what, following others – Margaret Archer and Andrew Sayer – I consider to be human universals: the proclivity for evaluation and normative judgement, the propensity to care about things, and the ability to flourish or suffer depending on whether what matters to us is diminished or enhanced. In bringing the concept of concern to the foreground of my analysis of ethical consumer practices and identities, I aim to do justice to the relationship of care that, as I aim to show in this book, goes a long way towards explaining human behaviour. I also hope to avoid reproducing "bland accounts of social life, in which it is difficult to assess the import of things for people" that Sayer (2011, p. 6) so convincingly calls us to break away from. In this book, I respond to his earnest appeal: through a sustained focus on the subjective meanings attached to consumer behaviour, I repeatedly demonstrate the connection between what people care about and which courses of action they take.

The value of the conceptual toolkit with which I approach the study of ethical consumption extends beyond its particular aptness for elucidating the inner workings of the mind of an ethical consumer. It also serves to reinforce and amplify the critical realist slant of the book, and is central to producing a more holistic, integrated, and balanced account of consumer engagement in ethical lifestyles. The concept of concern accentuates the relationship between human beings and the social world, unlike many other concepts that are widely employed in the explanations of human behaviour and that have a built-in bias towards either individual agents or social structure. In an illuminating account of *Why Things Matter to People*, Sayer (2011, p. 2) conveys this point:

> Concepts such as "preferences", "self-interest" or "values" fail to do justice to such matters, particularly with regard to their social character and connection to events and social relations, and their emotional force. Similarly, concepts such as convention, habit, discourses, socialisation, reciprocity, exchange, discipline, power and a host of others are useful for external description but can easily allow us to miss people's first person evaluative relation to the world and the force of their evaluations.

The concept of concern, however, enables social analysis to avoid over-emphasising the personal at the expense of the social, or vice versa, and to bridge the two levels in a subtle yet unmistakable way. This is because it accommodates both the subjectivity of agents – their being concerned about a particular thing out of an infinite number of things they could potentially be concerned about, and the objectivity of the external world – there being something real out there that arouses concerns, such as climate change, animal cruelty, or violation of human rights, to take examples from an ethical consumer's moral horizon. When we talk about concerns, we inevitably refer both to their subjective and objective component: a person, or a group of people, holding the concern along with emotions, intentions, and commitments engendered by it, *and* the objectively existing situations, circumstances, and relations to which the concern in question is linked. Due to this two-level reach, the concept of concern supports and extends the critical realist perspective from which I approach my enquiry, for it balances the focus on the intrinsic propensity of human beings to shape their lives around what matters to them with the emphasis on the embeddedness of subjective commitments and life projects in external reality.

The concept of reflexivity, another pillar forming the foundation of the theoretical framework developed in this book, has the same ontological derivatives: construed as an innate human capacity to consider oneself in relation to the surrounding contexts and vice versa (Archer, 2007), it effectively bridges structure and agency while underscoring their irreducible distinctiveness (for the very idea of someone reflexively deliberating upon something is predicated on the assumption of the ontological space between the subject and the object of deliberations). Allied to the idea of the internal conversation, the concept of reflexivity is indispensable to the representation of individuals as beings whose relationship to the world is one of concern; as Archer contends, it is our reflexive capacity, realised through the medium of the inner self-dialogue, that enables us to discern, evaluate, and appropriate particular concerns as well as decide the best way to address them within the specific contexts in which we are placed. Concerns, therefore, are conceived and nurtured by *a ménage à trois*: agents with an intrinsic propensity to value and care about things, objectively real events and circumstances which give rise to agential concerns, and reflexivity which begets and sustains the relationship of care between human beings and the surrounding world.

By revealing the fundamental forces at play in creating and shaping ethical consumption and highlighting their two-way interaction, my conceptual framework helps to break through the limits of traditional perspectives on consumer behaviour, oscillating between the agency-focused (wherein the individual consumer is the key author and governor of consumption) and socio-centric (in which consumption phenomena are created and shaped by social structures) outlooks. This contributes towards a key aim I pursue in this book, namely, to expose and redress the ontological and analytical biases that reside in the field of consumer research, and to encourage a more integrated and balanced approach to conceptualising, studying, and explaining consumer behaviour – an endeavour abetted and aided by the critical realist approach taken in this book. The study of ethical consumption lends itself to the application of critical realism as it provides an opportunity to explore connections between consuming agents, their capacities and liabilities, and objective reality – the social, natural, and practical contexts within which consumption takes place and whose properties and powers have causal effects on consumer conduct. On the one hand, we have individuals who actively, consciously, and creatively engage in alternative consumer behaviour because they made a commitment to a certain moral cause and recognise the sphere of consumption as an area of life in which they can effectively express this commitment in practice. On the other hand, the objective conditions in which aspiring ethical consumers find themselves exert a continuous influence on what, where, and how they consume and, to a significant degree, determine whether and to what extent their deeply held moral principles can be lived out and acted upon. One might be genuinely committed to the values and aims of the fair-trade movement but be cut off from the mainstream provisioning of fair-trade goods; or have a yearning desire to only eat local, yet live in a climate where such consumption routine is simply not viable; or passionately want to adopt a vegan lifestyle, but be suppressed by the socio-cultural norms and traditions surrounding eating behaviour in the given locale. What this tells us is that individual agency is always externally conditioned; its enactment never takes place outside of specific objective constraints. Likewise, ethical consumer identities are not freely appropriated; rather, they are attempted, negotiated, and liable to variation, transformation, and change under the influence of a multitude of subjective (personal concerns, needs, and desires directly affecting the extent and strength of individuals' commitments to ethical living) and objective (socio-cultural, political, and economic contexts) factors. Critical realism with its distinctive emphasis on and sensitivity towards a continuous interplay between the personal (agency) and the social (structure) provides an effective approach to explore how these subjective and objective elements interact with each other to create and define ethical consumer practices and identities.

While this book focuses on ethical food consumption, it raises questions about the sources and determinants of consumption processes and phenomena more broadly. Moving away from the myopia of one-sided views on consumer behaviour, this book reveals that consumers are neither absolutely free, nor are they completely constrained. Drawing insights from an empirical study with self-defined ethical food consumers, I will argue that it is only by acknowledging the key role of agential subjectivity and structural objectivity in shaping which courses of action individuals take, and hence what kind of persons they become, that we can achieve a true understanding of how –through which inner workings and under which external conditions – individuals develop and enact particular consumer identities. Through a causal analysis of individual commitments to ethical eating, this book demonstrates the benefits of applying critical realism in sociological investigations of consumption and argues the case for its primacy over one-dimensional frameworks in the battle to truly understand the reasons for ethical consumer behaviour.

Clarification of terms

Before letting the story of an ethical food consumer unfold, I would like to clarify some important terms which will be used repeatedly throughout the book and which, I feel, are liable to misinterpretation. The key term that calls for elucidation is, of course, "ethical consumption" itself. It is used to describe the phenomenon for which many other terms exist and are widely employed (mostly interchangeably, albeit at times with ambiguous distinctions) in the academic, media, and public discourses, such as green consumption, sustainable consumption, responsible, mindful, or conscious consumption, political consumption, critical consumerism, among others. In this book, I will refer to all of these as "ethical consumption".

There is no one clear or unified definition for ethical consumption, but it is commonly understood as a range of consumption choices, practices, and activities that are informed by individuals' morals, that is their understanding of what is right or wrong with respect to others, and usually oriented towards the needs of the natural environment, animals, and humans. Most scholarly attempts at defining the phenomenon link it closely to market contexts and shopping practices: Micheletti (2003, p. 2), for instance, defines political consumption as "actions by people who make choices among producers and products with the goal of changing objectionable institutional or market practices". Yet, ethical consumption encompasses a wide range of more subtle practices and activities that cannot be reduced to purchasing ethical products in the marketplace, and shopping is often not the only, or even primary, way in which people register their support for ethical consumption. Devinney, Auger, and Eckhardt (2010, p. 9) go as far as to claim that "the notion of ethical consumerism is too broad in its definition, too loose in its operationalisation, and too moralistic in its stance to be anything other than a myth". While I agree that ethical consumption is not a definitive concept, as someone approaching the question from a critical realist perspective I conceive of it as a real, multidimensional, and complex phenomenon which accommodates many varied interpretations and meanings and is presented by an array of practices, acts, and activities performed by socially situated and contextually circumscribed agents. Moving roughly along the same lines, Barnett, Cloke, Clarke, and Malpass (2005, p. 29) define ethical consumption as "any practice of consumption in which explicitly registering commitment to distant or absent others is an important dimension of the meaning of activity of the actors involved". The search for the ontological truth renders Barnett et al.'s definition noteworthy, for their statement captures the ontology of the phenomenon or, in other words, "what there is" to ethical consumption: an activity that is objectively taking place, its underlying intentions and motives, and a reflexive, creative, purposeful agent – the ultimate author of the activity and the sole central source of the subjective meanings invested in.

Yet, an important correction is warranted here: Barnett et al.'s pronounced emphasis on *distant* or *absent* others as the key focus of agential commitments is not only unnecessary, but altogether mistaken since, as we know from experience and scholarly accounts, ethical consumption practices are just as likely to be oriented towards those that are "closer to home". To understand consumer ethics solely as the ethics of distance, that is as an expression of care and regard towards those who are absent from the "here and now" is to leave out of view a whole range of other forms of ethical consumption, namely, those which exhibit the ethics of closeness. Local consumption, popularised by the Italian "slow food" movement, and growing availability of local vegetable box schemes (Littler, 2011), is often celebrated as the most ethical food choice which, as Adams and Raisborough point out, "works to disrupt any formulation linking the 'good choices' here with the livelihood of a producer 'over there' – 'distant or absent others'" (2010, p. 271). This book will provide further evidence to support this conclusion: whilst exploring consumers' conceptions and enactments of food ethics, I will note how for some protecting those who are closest, be it local producers, family members, or friends, was the highest moral priority and the key guiding principle of their commitment to ethical practices. Barnett et al.'s definition of ethical consumption in explicitly spatial terms seems even less appropriate in light of the authors' own acknowledgement of the excessive academic focus on the (arguable) causal relationships between physical distance, consumer knowledge, and the sense of moral responsibility which, as they note, "tends to underplay a range of other considerations that might play a role in shaping people's dispositions towards others and the world around them" (Barnett et al., 2005, p. 25). In this light, I feel it would be more appropriate to describe ethical consumption as "any practice of consumption in which explicitly registering commitment to others is an important dimension of the meaning of activity of the actors involved", a refined version of Barnett et al.'s definition, which is sufficiently specific, yet not suffocatingly prescriptive.

Another important term that re-emerges throughout this book is "moral". I use the words "morality" and "moral" to refer to the principles of right and wrong behaviour or, to introduce a more formal definition, "the internalized norms, values, principles and attitudes we live by in relation to other people" (Lindseth & Norberg, 2004, p. 145). Philosophers commonly draw distinctions of various degrees of sharpness between morality and ethics: Puri and Treasaden (2009, p. 1,223), for instance, propose that "ethics is the science of the philosophy of morals, and morals is the practice or enactment of ethics". In conversations with lay people, however, the dividing line between the two terms becomes very blurred, often disappearing altogether: my research participants, for instance, used the words "moral" and "ethical" interchangeably when discussing the issues of right and wrong, good and bad, virtue and vice. Given that for the study subjects the meanings of morals and ethics clearly overlap, I prefer to avoid the unhelpfully restrictive and often confusing ways of distinguishing between the two notions and, following the lead of other commentators on the subject, such as Andrew Sayer (2011) and Sam Harris (2010), will use them synonymously.

Another requisite clarification to be made concerns my use of the word "mind". Given the book's focus on exploring the emotional and mental workings underpinning the production of the ethical consumer identity, I deem it essential to guard against the narrow interpretation of the term as referring exclusively to human faculty of rationality or reason. I use it in its broader and, notably, primary sense to mean "the element of a person that enables them to be aware of the world and their experiences, to think, and *to feel*; the faculty of consciousness and thought" (Mind, n.d., my italics). Finally, when talking about subjective meanings, I use the term "meaning" to refer to "intention, cognition, affect, belief, evaluation, and anything else that could be encompassed in what is broadly termed the 'participants' perspective'" (Creswell, 2012, pp. 137–138). Looking through a critical realist lens, I treat these intentions, cognitions, beliefs, etc. as ontologically subjective, that is, existing only when and as experienced by an agent, but objectively real mental processes and phenomena that play a key role in defining individual and social outcomes.

I outline these nuances of meaning to prepare the readers for the realist account of ethical consumption that I present in this book and which, I hope, will yield a deeper understanding of ethical consumer behaviour through shedding light on the essential agential and structural properties, powers, and capacities and the particular commitments, practices, and identities emanating from them.

The scope and structure of the book

My intention in this book is to advance theoretical and empirical understanding of ethical consumer practices and identities, and to unlock the untapped potential of critical realism to guide research on consumption phenomena. Among the different forms of ethical consumption, I chose to focus on ethical eating due to the close relationship between food and identities – a connection which raises no doubts among experts in the field (Chapter 1 reviews how this relationship has been theorised and exemplified in sociological literature). Moreover, food consumption encompasses a wide range of ethical issues, including social justice and human rights, environmental and planetary wellbeing, and animal welfare, and thus can provide insights into a wide range of moral concerns underpinning consumers' adoption of ethical practices. At the same time, other types of consumer action – both positive, such as recycling, eco-travel, eco-fashion, and negative, such as boycotting – represent further avenues for exploring the issues of identity and selfhood surrounding consumer engagement in ethical practices. It is my hope that this book will provide a point of reference to elaborate on other relevant areas and extend the sociological debate on the underlying causality of ethical consumption behaviour.

Exploring ethical consumer identities is an intellectual puzzle and a compound research exercise which poses the need to examine the different phases, however vague and elusively demarcated they may be, that individuals go through as they progress towards attaining the desired self, and the key forces – agential capacities and structural powers – that inspire, enable, and shape this complex multi-level process every step of the way. It is this journey, spread across eight chapters, that this book invites the readers to embark on.

Chapter 1 offers a critical review of the key theoretical perspectives, agency-focused and socio-centric, that for the last several decades have been dominating sociological research on consumption in general and ethical consumption in particular. While acknowledging the contributions of these approaches to consumer studies as being of lasting value, I argue that empirical research informed by either of these two frameworks inevitably leads to a one-dimensional view of consumption and fails to achieve a nuanced understanding of both individual and social aspects of consumption processes and phenomena. I highlight the key ontological and methodological assumptions of both these perspectives and explain how they preclude a fuller understanding of the ways in which consumer practices are moulded and shaped. In light of this analysis, I argue for the need to produce a unified account of consumer behaviour which would match the complementary strengths and weaknesses of the agency-focused and socio-centric approaches. It is here that the book begins to demonstrate the benefits of critical realism for developing a more comprehensive and nuanced understanding of ethical consumption and consumer behaviour more broadly.

Building upon the preceding discussion, Chapter 2 foregrounds the relevance of critical realism for the field of consumer studies. First, I present a realist account of identity proposed by Margaret Archer and introduce its central concepts, including reflexivity, internal conversation, and ultimate concern. I draw on existing philosophical and sociological knowledge to establish the ontological status of reflexivity as a personal emergent power and a fundamental feature of personhood. Next, I lay out the critical realist ontology, outline its conceptual and analytical achievements, and discuss their implications for sociological investigations of consumption phenomena. I argue that the morphogenetic approach with its unwavering commitment to ontological realism and analytical dualism provides an effective theoretical and methodological framework for analysing consumer behaviour. More specifically, I emphasise stratified reality, pre-existence of social forms, and causal efficacy of both agents and structure as necessary preconditions for exploring the complex ensemble of individual and systemic powers which motivate, inform, and define consumption in general and ethical consumption in particular.

Chapter 3 places ethical consumption within the broad framework of reflexivity and, more specifically, reflexive construction of identity. My aim in this chapter is to propose a mechanism to account for the inner psychological process which brings consumers' ethical self into being. I apply Archer's theory of identity formation to delineate how through reflexive scrutiny of their subjective concerns and objective contexts individuals arrive at the decision to commit themselves to alternative – more environmentally, socially, morally responsible – ways of consuming. To this account, I bring insights from Coff's (2006) work on food ethics to advance an explanation as to how people become sensitised to concerns about the ethics of consumption in the first place. Through such theoretical integration, I reconstruct the chain of causal processes and interactions underlying the production of the ethical consumer identity. Grounded in critical realism, my account expounds the process and mechanism of formation of consumers' ethical self in a way that allows acknowledging and exploring the role of both agential subjectivity (human capacity for reflexivity, creativity, and intentionality) and structural objectivity (enabling and constraining properties of external reality) in shaping what people do and who they become.

Chapter 4 presents and discusses the research methodology underpinning my study on self-perceived ethical consumers. Its main purpose is to provide an overview of my chosen research strategy and techniques and situate them in the context of my philosophical position. I explain how my commitment to realist ontology and interpretivist epistemology shaped my approach to data production and analysis. The chapter considers in detail the problems inherent in my chosen research tools and the issues confronting the production of trustworthy and credible findings in interpretive research. I describe the steps that I took to harness the potential and mitigate the weaknesses of my chosen methods of investigation. Finally, I provide a personal account of self-reflexive enquiry in qualitative research: I highlight the challenges and opportunities presented by my positionality and subjectivity – as a researcher, an individual, and a consumer – and describe my endeavours to neutralise the impact of the self on the research process and outcomes.

The second part of the book integrates my theoretical arguments with empirical research. It opens with Chapter 5, which introduces my research participants – ten self-identified ethical consumers who volunteered to share their personal life and consumption stories. By means of individual vignettes – short stories generated from casual chats, informal discussions, and recorded interviews with respondents – I narrate each participant's personal background, ethical consumption practices, and their underlying concerns, as well as personal views on what being an ethical consumer entails and what kind of challenges and rewards it involves. These brief portraits make important references to participants' personalities and life contexts and are key to understanding their pathways to ethical consumption. This chapter intends to help the readers to better grasp the empirical examples used in the book by locating them within the context of the respondents' lives as individuals and consumers. It thus provides the essential backdrop against which my analysis of ethical consumer practices and identities can unfold.

In Chapters 6, 7, and 8, I present the findings from my research on ethical consumers to provide empirical support for the theoretical ideas and claims set forth in the first part of the book. Chapter 6 seeks to demonstrate the empirical relevance and explanatory force of my proposed theoretical account of the ethical consumer identity. By analysing participants' narratives of their mental and emotional journeys towards ethical consumption, I empirically re-construct the inner psychological process through which ethical consumer identity comes into being. I trace participants' ethical concerns back to their "glimpsed" experiences of consumption-related ethical issues and demonstrate their role in triggering the reflexive workings leading to the decision to commit to an ethical lifestyle. Further, I examine participants' internal conversations to illustrate the inextricable relations between ethical consumer concerns, commitments, practices, and identities. While demonstrating the central part played by reflexivity in the production of consumers' ethical self, I simultaneously underscore the inherent fallibility of human reflexivity and the inevitable constraints placed by objective reality on consumers' pursuits of desired identities.

In Chapter 7, the focus of analysis shifts from the challenges of becoming to the complexities of being an ethical consumer. The key argument here is that the attainment of the ethical consumer identity is not a self-sustaining achievement, but one that requires ongoing maintenance by active, reflexive, and intentional agents. I draw on the biographical narratives elicited from respondents to reveal the contextual embeddedness of consumption practices and demonstrate how the properties of objective reality exert direct causal effects on consumer behaviour. Next, I present evidence emphasising the individual's capacity to actively interact with, creatively respond to, and reflexively negotiate structural influences, both constraining and habilitating, to ensure that their identity-defining ethical projects come to fruition. Here again attention is drawn to the key role of reflexivity in enabling ethical consumers to continuously monitor the self, its subjective concerns, and objective contexts and sustain a fulfilling and feasible life. This chapter also reveals what it means for individuals to not merely enact ethical consumption, but to deeply identify with it. Based upon my analysis of respondents' self-narratives, I argue that for a morally concerned and committed consumer, maintaining coherent and stable ethical practices is key to preserving the sense of personal integrity, continuity, and self-worth. I corroborate my argument by exemplifying the identity implications of contradictions and inconsistencies exhibited by self-defined ethical consumers and identifying a set of ideational strategies which they use to defend their moral self-image.

Finally, in Chapter 8 the focus shifts from ethical consumers' inner to their outer selves. The key aim of this final chapter is to provide an account of social identity formation in ethical consumers. Here I present Archer's theory of social identity, which I bring into dialogue with an empirical analysis of the links and interactions between consumers' ethical practices and their social lives and relationships. I demonstrate how commitment to ethical consumption extends beyond the individuals' self-concepts to lay claim to their social selves, and how sociality in turn affects the continuity and consistency of consumers' ethical projects. Further, I analyse how and with what implications for their personal and social identities consumers arbitrate between ethical and social concerns. This chapter therefore identifies, explores, and empirically illustrates the underlying mechanism that leads to the formation of the ethical consumer persona and accounts for the varying degrees of its visibility in the life of an individual.

Together, these three last chapters provide a detailed and integrated account of the process of becoming and being an ethical food consumer. In recounting participants' stories, I do not aim to reconstruct their biographies; rather, my goal is to bring into the spotlight those experiences and events from their lives as unique persons and consumers in which the real causal processes underlying the production of the ethical consumer identity are most effectively captured. Yet, one important story will have been told in the end: the story of an ethical food consumer – a human being endowed with various powers and capacities and burdened with a multitude of concerns, a moral agent in pursuit of his or her precious and authentic life project, and a social actor occupying a wide range of actively chosen as well as involuntarily imposed positions and roles, each with its accompanying obligations and duties. This story, in which the leading parts will be assigned not to particular individuals, but to abstract ideas, such as reflexivity, identity, and concerns, as well as broader concepts of structure and agency, will take us from the concrete realm of personal experiences to the higher levels of abstraction where a true understanding of the real causal mechanism of the ethical consumer phenomenon can be obtained.

The conclusion highlights what I see as the key contributions of this book to our current understanding of ethical consumers and their behaviour. Here I draw attention to the inevitable limitations and blind spots of my study on self-defined ethical consumers and provide recommendations for future research which, I suggest, should proceed more confidently and systematically towards a truly dialectical perspective on consumer practices and identities.

References

Archer, M. (2007). *Making Our Way through the World*. Cambridge, UK: Cambridge University Press.

Askegaard, S., & Kjeldgaard, D. (2002). The water fish swim in? Relations between culture and marketing in the age of globalization. In K. Thorbjørn, S. Askegaard, & N. Jørgensen (Eds.), *Perspectives on Marketing Relationships* (pp. 13–35). Copenhagen: Thomson.

Barnett, C., Cloke, P., Clarke, N., & Malpass, A. (2005). Consuming ethics: articulating the subjects and spaces of ethical consumption. *Antipode, 37*(1), 23–45.

Belk, R. W. (1988). Possessions and the extended self. *Journal of Consumer Research, 15*(2), 139–168.

Coff, C. (2006). *The Taste for Ethics: An Ethic of Food Consumption*. Dordrecht: Springer.

Coskuner-Balli, G., & Thompson, C. (2009). Legitimatizing an emergent social identity through marketplace performances. In McGill, A., & Shavitt, S. (Eds.), *Advances in Consumer Research* (Vol. 36, pp. 135–138). Duluth, MN: Association for Consumer Research.

Coulter, R. A., Price, L. L., & Feick, L. (2003). Rethinking the origins of involvement and brand commitment: insights from postsocialist central Europe. *Journal of Consumer Research, 30*(2), 151–169.

Creswell, J. (2012). *Qualitative Inquiry and Research Design*. London: Sage Publications.

Devinney, T., Auger, P., & Eckhardt, G. (2010). *The Myth of the Ethical Consumer*. Cambridge, UK: Cambridge University Press.

Dobscha, S., & Ozanne, J. L. (2001). An ecofeminist analysis of environmentally sensitive women using qualitative methodology: the emancipatory potential of an ecological life. *Journal of Public Policy & Marketing, 20*(2), 201–214.

Gabriel, Y., & Lang, T. (2006). *The Unmanageable Consumer*. Thousand Oaks, CA: Sage Publications.

Gopaldas, A. (2014). Marketplace sentiments. *Journal of Consumer Research, 41*(4), 995–1014.

Harris, S. (2010). *The Moral Landscape: How Science Can Determine Human Values*. New York, NY: Free Press.

Holt, D. B. (1997). Poststructuralist lifestyle analysis: conceptualizing the social patterning of consumption in postmodernity. *Journal of Consumer Research, 23*(4), 326–350.

Holt, D. B. (2002). Why do brands cause trouble? A dialectical theory of consumer culture and branding. *Journal of Consumer Research, 29*(1), 70–90.

Kozinets, R. (2002). Can consumers escape the market? Emancipatory illuminations from Burning Man. *Journal of Consumer Research, 29*(1), 20–38.

Lindseth, A., & Norberg, A. (2004). A phenomenological hermeneutical method for researching lived experience. *Scandinavian Journal of Caring Sciences, 18*(2), 145–153.

Littler, J. (2011). What's wrong with ethical consumption? In Lewis, T. & Potter, T. (Eds.), *Ethical Consumption: A Critical Introduction* (pp. 27–39). London: Routledge.

Micheletti, M. (2003). *Political Virtue and Shopping*. New York, NY: Palgrave Macmillan.

Miller, D. (1998). *A Theory of Shopping*. Cambridge, UK: Polity Press.

Mind [Def. 1] (n.d.). In Oxford Dictionaries Online. Retrieved 2 March 2017, from http://www.oxforddictionaries.com/definition/english/mind

Muniz, A. M., & O'Guinn, T. C. (2001). Brand community. *Journal of Consumer Research, 27*(4), 412–432.

Oswald, L. R. (1999). Culture swapping: consumption and the ethnogenesis of middle-class Haitian immigrants. *Journal of Consumer Research, 25*(4), 303–318.

Puri, B., & Treasaden, I. (2009). *Psychiatry: An Evidence-based Text*. London, UK: Taylor & Francis.

Sayer, A. (2011). *Why Things Matter to People*. Cambridge, UK: Cambridge University Press.

Wallendorf, M. (2001). Literally literacy. *Journal of Consumer Research, 27*(4), 505–511.

Woodward, I. (2003). Divergent narratives in the imagining of the home amongst middle-class consumers: aesthetics, comfort and the symbolic boundaries of self and home. *Journal of Sociology, 39*(4), 391–412.

Woodward, I. (2007). *Understanding Material Culture*. London: Sage Publications.

Part I

Theorising the ethical consumer

1 Analysing consumption

Towards an integrated approach

An approach which integrates social influences and scope for reflexivity and responsibility can explain things which neither of these one-sided theories can.

(Sayer, 2011, p. 56)

In much of the sociological literature on ethical consumption, the display of alternative consumer positions and attitudes has been conceptualised in terms of collective action in pursuit of political and social progress. Within the stream of research interpreting ethical consumption as a form of political participation and governance, the work of Micheletti (2010, 2011) has been especially influential, but many other scholars in the field of consumer studies have approached the analysis of ethical consumer behaviour from the same conceptual angle (e.g. Boström & Klintman, 2008; Clarke, Barnett, Cloke & Malpass, 2007). Representations of ethical consumption as a means of political engagement and a vehicle for social change are based on an implicit assumption that consumers' adoption of ethical lifestyles is driven largely by practical goals, such as raising awareness of the deficiencies of modern production systems and driving structural changes in agriculture and industry. Reflecting the growing interest in ethical consumers as citizens and political agents, the focus of academic enquiry has been predominantly on external manifestations of the "consumer self" and its effectiveness in enacting social change. Meanwhile, aspects of individual engagement in ethical consumption have remained in the shadows, with very few concerted efforts being directed towards producing an effective account of the subjective meanings and personal motives invested in ethical consumer behaviour. In 2001, Tallontire, Rentsendorj, and Blowfield undertook a wide-ranging review of academic literature on fair trade, which revealed a glaring gap in the contemporary understanding of the meanings of ethical purchases for individual consumers and the ways these meanings translate into actions, highlighting the need for more exploration into this area.

Since then, however, there has been an observable proliferation of research aimed at recognising and exploring the implicit and explicit motivations, intentions, aspirations and goals attached to ethical consumer choices. In this chapter, I will situate this burgeoning stream of literature vis-à-vis the prevailing theoretical approaches to consumption that emerged and developed in the last several decades and whose core presuppositions have been informing empirical investigations of consumer behaviour. My aim is to critically review these frameworks in order to identify and expose their ontological and analytical biases, which continue to inhibit a comprehensive understanding of consumption phenomena at both the individual and social levels and, building upon this critique, argue for the benefits of critical realism for developing a much more complete, balanced and nuanced perspective on ethical consumption and consumer behaviour more broadly.

A view on consumption: lessons from the past, directions for the future

Since the 1980s, there has been a considerable increase in the scholarly attention to the subject of consumption. Among sociologists, a once dominant theoretical view of consumer habits as a direct reflection of material circumstances and class positions has gradually lost its appeal; a more nuanced understanding of consumption as shaped by a wide range of individual and social forces has arisen instead (Warde, 1997). Inspired by this new, more extensive understanding of the motives and antecedents of consumer choice, various perspectives on consumer behaviour emerged which have placed the focus of conceptual and analytical concern at different locations along the structure-agency spectrum, depending on whether society or the individual is seen as the ultimate author and source of consumption practices. At one end of this spectrum are theoretical views that take the consumer to be the prime mover of practices and a chief focus of scientific investigation, while on the other side are socio-centric approaches within which consumers are conceived of as merely bearers of practices, and the scientific interest shifts towards the social roots of consumption behaviour and the wider societal contexts in which it takes place.

Among agency-focused frameworks, the theorisation of consumers as identity-seeking, meaning-creating individuals engaged in a continuous reflexive process of constructing a coherent self through the creative appropriation of a range of commodities has been highly influential. The idea that consumption serves as the main medium in which the reflexive project of the self unfolds, has been substantiated by some of the most influential thinkers in the field of consumer studies, as demonstrated by the following quote from Alan Warde (1997, p. 68):

> today, people define themselves through the messages they transmit to others via the goods and practices that they possess and display. They manipulate or manage appearances, thereby creating and sustaining a 'self-identity'.

Consumer Culture Theory has played a vital role in inspiring a systematic and extensive enquiry into the part played by consumption in identity creation and communication and promoting the image of the active, freely choosing consumer reflexively engaging with mythic and symbolic resources circulating within the post-modern marketplace. Arnould and Thompson's (2005) synthesising review of two decades of research on the symbolic, socio-cultural, experiential, and ideological aspects of consumption contains numerous examples of studies theorising and empirically demonstrating the links between individuals' identity projects and consumption behaviours. The view of consumption as an arena of reflexive self-production and consumers as active agents continuously negotiating their identities through a complex variety of product choices has penetrated into sociological thinking about eating and food. On the one hand, associations between what people eat and their personal and social identities have been claimed (Fischler, 1988; Lang & Heasman, 2004; Warde, 1997) and exemplified in research: Warde's (1997) study of culinary recipes in popular women's magazines, Goodman's (2004) analysis of the contemporary nature of fair trade, and Diner's (2001) investigation of food practices of three distinct migrant groups in America all finely argue for the symbolic role and identity value of food. On the other hand, the idea of reflexivity has been introduced into sociological accounts of eating patterns in post-traditional societies to compensate for the "decline in 'the spirit of discipline'" (Warde, 1997, p. 13) in the domain of food consumption. It has been argued that in a world where people are no longer embedded in traditional social contexts and no longer belong to familiar collectivities, the questions of what, when, and how to eat are increasingly a matter of individual rather than collective decisions (Fischler, 1980). In the absence of a social and cultural framework for eating habits, so the argument goes, individuals lack the usual reassurance about their dietary behaviour:

> Denied is the sense of comfort and security that can be derived from knowing that our tastes and preferences, even in the humble field of food, are endorsed and shared by others, whom we respect and with whom we consider we belong.
>
> (Warde, 1997, p. 173)

In such conditions reflexivity takes over from traditions to provide guidelines for appropriate eating practices, and a reflexive food consumer – the one who exhibits a "broader sense of agency in the realm of consumption choices, reflected in knowledge-seeking, evaluation, and discernment" (Guthman, 2002, p. 299) – emerges.

The requirement to be reflexive has intensified as a result of processes caused by rapid, radical transformation of the global food environment. A succession of safety scandals plaguing modern-day food industry (well exemplified by the salmonella controversy of 1988, the Alar scare of 1989, the BSE crisis of 1996, the E. coli outbreak of 2011, and the horsemeat scandal of 2013) and unprecedented advances in production technologies ceaselessly fuel public thinking about food in terms of danger and risk, which increasingly self-reliant and autonomous consumers have to negotiate on their own. The profile of "discerning food consumers" (Murdoch & Miele, 1999, p. 469) has been further rising in the light of mounting evidence and a growing recognition of the adverse effects of the modern food system on our physical, societal, economic, and environmental well-being

(Fraj & Martinez, 2007; Lang, Barling, & Caraher, 2009). This is symptomatic of Beck's risk society, wherein the notion of risk is systematically generated and nurtured by "hazards and insecurities induced and introduced by modernisation itself" (Beck, 1992, p. 21). Concomitantly, studies began to appear suggesting that people are progressively incorporating reflexivity in their daily consumption decisions (Arvola et al., 2008; Torjusen, Lieblein, Wandel, & Francis, 2001). The same conclusion wraps up Hilton's analysis of a centuries-long discourse on consumption, indicating that "an increasing number of consumers are beginning to think more closely and more often about the basis of their own comfort" (2004, p. 119).

The figure of a reflexive, identity-pursuing consumer has consequently established a presence in sociological accounts of ethical consumer behaviour (see, e.g., Adams & Raisborough, 2008; Barnett, Cloke, Clarke, & Malpass, 2005; Cherrier, 2006; Gabriel & Lang, 2006; Halkier, 2001; Micheletti, 2003). Examples of studies emphasising the links between various forms of ethical consumption and individual as well as collective identities abound. Shaw's (2007, p. 141) investigation of boycotting behaviour describes a group of consumers for whom the display of alternative consumer attitudes and positions was an important way of "marking your own identity". Shaw and Shiu's (2003) earlier enquiry into the factors influencing ethical choice and Newholm's (2005) research on consumer engagement in responsible shopping both argue that the integrity of personal identity is a key motive behind ethical consumer behaviour. Taking the identity theme further still, sociological research started to supply commentary on the potential of ethical consumption to serve not only as a tool for self-construction, but also as a mechanism of self-control whereby "individuals in the act of constructing and reconstructing their own biographies monitor their own behaviour and thereby, at least half-consciously, discipline themselves with a view to self-improvement" (Warde, 1997, p. 93). Barnett et al. (2005, p. 29) epitomise this idea in the concept of "moral selving", which refers to the process of creating and displaying different forms of selfhood through engagement in alternative consumption practices.

The view of ethical shopping as a means of cultivation of a better self through morally responsible choices finds considerable support in empirical research. Kozinets and Handelman's (1998) study of boycotts, for example, highlights the powerful individualizing and morally transforming the potential of boycotting behaviour which consumers tap into to define "a personal morality that has 'evolved' beyond hedonistic commercial interests". Ethical consumption, authors argue, creates opportunities for the activation of such values as compassion, care, reciprocity, and responsibility, through which consumers can materialise their ideal self. Likewise, Moisander and Pesonen (2002, p. 330) interpret ethical consumer behaviour as a mode of self-formation that involves "a permanent questioning and reinventing of the self". In their study of environmentalism, the authors discuss how the practice of green living allows individuals the opportunity to re-invent themselves as moral subjects as opposed to materialistic consumers, and how acts of ethical consumption can be used as elements in the "politics of the self". Similar conclusions emerge from Cherrier's (2006, p. 520) analysis of consumer use of eco-friendly shopping bags revealing the role of ethical product choices in shaping a person's view of herself as "a recycler, a green voter, an environmentally conscious consumer or an ethical citizen". These studies offer an empirical record of the potential of ethical consumption not only to tell "the story of who we are", but also to fulfil the "fantasy of what we wish to be like" (Gabriel & Lang, 2006, p. 94). The interpretation of ethical consumption as a means for moral self-enhancement clearly presupposes agency, reflected in the ability of consumers to resist and refuse materialist subjectivities imposed by the dominant consumer culture and imagine, create, and promote alternative forms of individuality.

Another stream of research has drawn out lessons for understanding the motives underlying ethical consumer behaviours from the argument that "we use consumption symbolically not only to create and sustain the self but also to locate us in society" (Wattanasuwan, 2005, p. 179). This line of thought interprets ethical shopping through Veblen's (2009) lens, namely as a form of "conspicuous consumption" aimed at projecting a higher social, economic, and cultural status through appropriation and display of commodities that confer particular attributes on those who possess them. The representation of ethical consumption as a strategy for social distinction rests on the assumption that being a responsible consumer presupposes certain levels of financial and cultural capital. In addition, Veblen's original view of expensiveness as a key product characteristic that provides the impression of social superiority has been re-thought to argue that "class is not just a matter of money" (Warde, 1997, p. 175), and that other symbols can offer channels for social distinction. The idea that values find material expression in goods itself is not new: in his classic study of the landmark Parisian department store, Le Bon Marché, Miller (1981) describes how the important bourgeois value of respectability was made "real" and "concrete" through a range of material goods, such as clothing and furnishing. Accordingly, some commentators have put forward the view that ethical products may be used as a proxy for personality traits – compassion, selflessness, kindness – that bring high-status rewards independently of financial success (Allison, 2009; Barnett et al., 2005). The potential of goods with ethical attributes to contribute to a desired self-presentation has been demonstrated through empirical research. The following quote from a participant of Shaw et al.'s (2005, p. 190) study of ethical shoppers illustrates the perceived status-enhancing effect of ethical choices:

> If you're putting Cafedirect [Cafedirect is a brand of fair trade coffee in the UK] in your trolley and driving around with it, then you're saying to other people I'm clever enough to know the difference between this and Nescafe.

A previously mentioned study of green shopping bag users by Cherrier (2006) gives another line of empirical evidence for the potential of ethical consumption to assist individuals in creating and managing their social image.

All aforementioned approaches to consumer behaviour are united by common conceptual leanings and analytical tendencies. First, they reflect a shift of academic focus from product features and attributes to the symbolic meaning and identity value, that is, the potential to create and communicate one's self-concept, of goods. More importantly, they share the view of consumers as active, interpretive, and intentional agents who solely author and freely perform consistently conscious and meaningful acts of consumption. Embedded within this perspective is the assumption that reflexivity acts as the key driving force and supreme determinant of consumer behaviour. Adams (2003) defines this approach as "the extended reflexivity thesis" (p. 222), characterised by the attribution of "a heightened, transforming level of reflexivity" (p. 221) to consuming agents engaged in continuous reflexive self-production.

The main contested features of such a perspective lie in its overemphasis on individual choice at the expense of acknowledging the role of structure in shaping the self and its practices and its positioning of reflexivity outside the particular social contexts in which it is exercised by agents. Both of these theoretical *faux pas* have attracted extensive and well-deserved criticism. The representation of identity as a project free from determination by external forces has been subject to unsympathetic scrutiny by thinkers outside and within the field of consumer studies. Tucker (1998, p. 208), for example, warns that "A strong self which heroically creates narratives of personal development in uncertain times ... gives short shrift to the structural and cultural factors still at work in fashioning the self".

Sassatelli (2007, p. 106) closely echoes the point whilst helpfully shifting the focus towards the consumption domain: "the ongoing constitution of a personal style draws on commodities whose trajectories consumers can never fully control and it is negotiated within various contexts, institutions and relations which both habilitate and constrain subjects". Cherrier too problematises the idea of an ethical consumer who "self-creates through will, operates freely in its own construction, and consciously chooses elements in the marketplace that meet its need for a meaningful or authentic identity" (2007, p. 322). In the context of this discussion, highly relevant and instructive is Trentmann's (2006) collection of essays exploring the making of the consumer in different social and economic contexts. The historical analyses of the evolution of a consumer as a social subject presented in the book clearly underscore the role of national institutions in creating and refashioning consumer roles and identities.

The idea of context-transcendent reflexivity has also come under sustained attack from socially attuned commentators. Archer (2007) is explicitly critical of the belief in unbounded reflexivity symptomatic of late-modernist theorisations of selfhood. Far from subverting the centrality of reflexivity to the construction of self and organisation of social life, Archer's argument nevertheless demands that the causal powers of social structures be acknowledged and their role in shaping agential answers to questions about "What to do? How to act? Who to be?" (Giddens, 1991, p. 70) be accounted for. While Archer's theory unequivocally places the reflexive process at the heart of identity formation, by no means does it invite us to think of human reflexivity as an unconstrained force flowing freely in an unstructured environment; to the contrary, consideration of the interaction between the causal powers of agents and those of social structures is strongly and zealously called for. Strong objections to the assumption of the unlimited scope and extent of reflexivity have also been voiced by Adams (2003), whose analysis of the late-modern accounts of identity draws attention to the lamentable tendency of contemporary social theory to overlook the social and cultural embeddedness of reflexivity and overstate its bearing upon the making of the modern identity. This criticism extends into a later work by Adams and Raisborough (2008), exploring how reflexivity manifests itself in ethical consumption. A critical analysis of the conceptual fit between consumption of fair trade and reflexive self-production once again accentuates the poverty of the extended reflexivity thesis which, in the words of the authors, precludes "an understanding of the specific and localized ways in which reflexivity emerges from a complex interface of socially and culturally stratified contexts, dynamic interpersonal relations and psychodynamics" (Adams & Raisborough, 2008, p. 1169).

Rational choice theory (RCT) denotes another agency-focused approach that has been widely applied in consumer research and that spilled over into the subject area of ethical consumption. Whilst sometimes classified as an offshoot of the extended reflexivity thesis (see Adams, 2003), rational choice perspective takes a markedly different stance with regard to the key goals and properties ascribed to consuming agents. The hallmark of RCT is its pronounced emphasis on rationality as the dominant human feature, hence the view of a social agent as a dispassionate, preference-driven, and goal-oriented actor making choices "on the basis of deliberate, systematic calculation of the maximum extent to which the ends can be met by using the inevitably scarce means" (Chang, 2014, p. 20).

In the field of ethical consumption research, the *homo economicus* model has become an inspiration for the interpretations of ethical consumer behaviour that dispense with the ideas of altruism, selflessness, and goodwill, with which voluntary adoption of ethical lifestyles is typically associated. From the viewpoint of RCT, consumer engagement in ethical practices is best construed as a form of self-pleasing behaviour on the part of a rational individual who does good not to *be* good but to *feel* good, that is, in a rationality-driven pursuit of his own self-interest. A prominent example of this line of thinking is Kate Soper's (2007, 2008) alternative hedonism thesis, which puts emphasis on the self-satisfying dimension of ethical consumption – the "sensual pleasures of consuming differently" (Soper, 2008, p. 577). Soper's account is grounded in the idea that modern consumer society is ultimately bound to throw its inhabitants into a state of profound dissatisfaction: "people are beginning to see the pleasures of affluence both as compromised by their negative effects and as pre-empting other enjoyments" (2009, p. 4).

Conversely, through engagement in ethical consumption, Soper argues, one can attain the material simplicity of life and in doing so reclaim the subtler forms of hedonist pleasures that have fallen prey to the dominant materialistic lifestyles. The ultimate rationale for consumer adoption of ethical practices, therefore, boils down to a pursuit of the life of pleasure, while reflexive engagement with environmental, social, and moral concerns is seen merely as a quest for "the self-massaging comfort of 'doing good'" (Lekakis, 2013, p. 78).

Viewed through a rationalist lens, ethical consumer choices appear void of an altruistic component and are best described as acts of selfish behaviour arising out of a rational desire to do good if doing good ranks high on the list of an agent's preferences. In the following quote, Archer (2000, p. 54) provides an insightful diagnosis of this condition: "*homo economicus* can have a taste for philanthropy, in which case it is the task of his reason to make him a well satisfied philanthropist, a cost-benefit effective benefactor and a philanthropic maximizer".

Tuning in to the alternative hedonism thesis, a range of accounts of ethical consumer behaviour has attempted to bring to the surface the self-interest presumably underlying individuals' engagement in ethical consumption. Arvola et al.'s (2008, p. 445) study of organic shoppers reports a connection between "positive self-enhancing feelings of 'doing the right thing'" anticipated by consumers and their intentions to buy organic. John, Klein, and Smith's (2002) research points out the "clean hand motivation" as a major driver of consumer boycotts. Investigations into the guiding motives of ethical shoppers by Cherrier (2006) and Shaw (2007) add more empirical evidence of the role of the "feel-good" factor in galvanising ethical consumer behaviours. More recently, a more nuanced and sophisticated understanding of the relationship between the self-interested and altruistic motives of ethical consumers has emerged: in her book on the politics of fair-trade consumption, Lekakis (2013, p. 78) interprets consumer involvement in coffee activism as a pursuit of a morally satisfying "state of equilibrium between the self-centred self (the hedonistic consumer who seeks 'the good life') and the self-governed self (the responsible, civically minded political consumer)".

Founded on the assumption of rational self-interest and guided by the concept of preference, such representations of consumer behaviour sit uneasily with claims about the inherently moral and value-laden nature of consumption reverberating in the works of various authors. Miller's (1998) year-long ethnographic study of shopping on a North London street offers an incisive account of the emotional, moral, and relational underpinnings of shopping practices. Contrary to what a rational choice theorist would assert, Miller's research tellingly demonstrates that even the most ordinary and routinised consumption involves handling complex and delicate moral issues and is best understood as a project about social relationships – those of care, commitment, responsibility, and love. Hilton's (2004) analysis of the evolution of moral discourses around consumption spanning the past three centuries provides equally compelling reasons for asserting that consideration of morality is central to understanding human consumption, both past and present. The argument is of particular relevance for the domain of food: the moral and ideological significance of cooking and eating practices has been widely acknowledged in sociological literature (Mennell, Murcott, & van Otterloo, 1992; Murcott, 1983; Warde, 1997). In a study of people's sources of culinary recipes, McKie and Wood (1992) highlight the cultural relevance of recipes and their role in setting social standards for cooking and eating behaviours. Likewise, Warde's (1997) study of culinary columns in women's magazines reveals the nuanced symbolism and powerful moral charge of day-to-day food choices.

It is, therefore, unsurprising that RCT with its flat denial of human normativity and emotionality is regarded with mistrust and doubt by sociologists of consumption, many of whom have expressed unreserved criticism of the framework's key postulates. Warde (2015, p. 121), for example, rejects the economic model of a man because of the simple fact that "people typically find within their activities both frustrations and satisfactions, anxieties and pleasures, not all of which are simple matters of calculation" – a claim which we know to be true both intuitively as well as experientially. Wilk too is strongly opposed to choice theorists' over-rationalised understanding of consumption which, he argues, "is in essence a moral matter, since it always and inevitably raises issues of fairness, self vs group interests, and immediate vs delayed gratification" (2001, p. 246), and hence cannot be reduced to mere calculations of losses and gains. Martha Starr's (2009) research concerned with explaining the drivers behind the rapid growth of ethical consumerism supplies further evidence that ideas of right and good versus wrong and bad feature prominently in people's decisions about the purchase and use of resources and goods. In a comparative study of the ethical wine industry in Australia and the United Kingdom, Paul Starr (2011, p. 137) highlights the failure of the rational choice approach to discern in consumer behaviour anything other than "the signaling of human demand" and to recognise that "consumers sometimes choose to relinquish their rational-sovereign, market-democratic role and make preference-decisions on non-market, even irrational grounds".

While these are important and valid objections, a realist project requires a more meticulous analysis of the underlying assumptions of the rational choice framework and a more nuanced critique thereof. The model of a social agent as a consistently rational and preference-driven chooser has a number of built-in ontological presuppositions that render RCT ill-suited for explicating human behaviour, including in the sphere of consumption. The representation of human morality as merely a part of the cost-benefit analysis of a narrowly self-interested actor who prefers that course of action which, alongside other utilities, also brings higher emotional rewards (Becker, 1996), leaves no room to accommodate such widespread sociocultural phenomena as altruism, benevolence, social solidarity, free-giving. The idea of human actions being pre-defined by a set of preferences that "are assumed to be given, current, complete, consistent and determining" (Archer, 2000, p. 68) denies agential interactions with objective reality any role in making us who we are and subverts the innate capacity of all people for reflexive deliberations upon the inner self and the outer world. This takes an unbearable toll on the essential human properties of emotionality, normativity, and reflexivity: devoid of the need to actively define and continuously reassess their concerns, subjects are left with their emotions untriggered, normativity unexercised, and the workings of the mind reduced to a cost-benefit analysis (Archer, 2007).

RCT's flat rejection of altruism becomes undeniably problematic when applied to ethical consumer behaviour which implies at least a degree of interest-free and self-sacrificing morality, as evinced by a growing number of people willingly foregoing their own convenience, leisure time, and material interests out of concern for the fate of "the other" – humans, animals, or the planet (think of those who give up their cars for the benefit of the environment, or spend yet another Sunday digging vegetable patches in a persistent effort to "grow their own", or pay significant price premiums for fair-trade goods). It is difficult to see how such behaviours can be reconciled with RCT's ontological assumptions of social atomism and individualistic, "rational-acquisitive reflexivity" (Donati & Archer, 2015, p. 278). As De Groot and Steg (2009) point out, the practice of ethical consumption requires foregoing gratification of one's own immediate desires and short-term interests for the sake of promoting the common good and hence must be at least partially founded in self-sacrificing morality. As such, it can only spring from an inter-subjective relational social ontology wherein agents are construed not as isolated individuals, but as parts of a system of interdependence character-ised by a growing interaction, reciprocity, and relationality. Likewise, reflexivity that engenders ethical actions cannot be merely individual; rather, it is relational, for it "reflects on the outcomes of social networks as products of relations rather than of individual acts" (Donati & Archer, 2015, p. 278).

An attempt has been made by rational choice theorists to explain away acts of charity, benevolence, and goodwill by rethinking the individual without conceding rationality as her dominant property. The refined model is that of a tripartite being consisting of a superior rational actor, a normative man introduced as a source of the sense of cooperation arising when common good is at stake, and an emotional man called upon when the expression of solidarity and collective action is needed for the sake of social stability or change (Flam, 2000). Archer (2000, p. 76) has spared me the task of exposing the ontological flaws of this model, in which

> The chain of rationality is not broken by the subsumption of action under normative expectations, because cultural dopery is avoided by asserting that the reasons for actions associated with a role, move an actor only when they are adopted as his own good reasons.

Archer advances several compelling objections to this theoretical configuration. Analytically, such a multi-layered model of a social actor makes the preferred focus of rational choice theorists on an individual as a basic unit of investigation difficult to sustain. Conceptually, it is hard to imagine by what means the three agents co-existing within a single human being can be kept hermetically com-partmentalised and, furthermore, harmoniously orchestrated so that they manifest themselves at appropriate places and times. Moreover and, perhaps, most impor-tantly, the construct is fundamentally flawed in that it incorporates the social into the individual: distribution of economic resources is narrowed down to personal budgets; social solidarity is explained away as merely an expression of a subjec-tive preference to team up; and subscription to social norms is construed as a rational pursuit of self-interest rather than a manifestation of a morally binding duty. "Can the social context really be disaggregated in this way?" (Archer, 2000, p. 67) and "in what recognisable sense are we still talking about 'the individual' when he or she has now been burdened with so many inalienable features of social reality?" (ibid.) are the ontological puzzles that RCT's recast of a human being leaves unresolved.

Finally, there are a number of ontological fallacies that rational choice theorists share with the proponents of the extended reflexivity thesis. The two frameworks are close theoretical allies: both subscribe to ontological and methodological individualism (where social reality is understood in terms of the aggregate outcomes of the motivations and actions of individual agents, and where the individual is used as the ultimate unit of analysis for empirical investigations), and both endorse the view of consumers as active and teleological decision-makers, operating in a highly individualistic and free-choice environment. The limitless rationality assumed in RCT parallels post-modernist belief in unbounded reflexivity: on both accounts, consumption practices are seen as the result of the free choices of consuming agents – identity-concerned and meaning-seeking individuals in one case; preference-driven and utility-maximizing actors in the other. Both approaches embody a neoliberal notion of consumers as knowledge-grounded agents choosing freely, whether rationally or reflexively, "how to be and how to act" (Giddens, 1994, p. 75), and both are fundamentally flawed, conceptually and analytically, in that they abstract consuming agents from their contexts and neglect the systemic and structuring influences of the social, political, and economic environments within which any act of consumption takes place. This persistent failure to take into account the contextual embeddedness of consumer experiences has invited a lot of criticism from structurally oriented scholars, emphasising that consumption amounts to a "complex economic, social and cultural set of practices" (Sassatelli, 2012, p. 236) and hence cannot be reduced to explanations derivable from facts about individuals, their properties and relations (see, e.g., Askegaard & Linnet, 2011; Hilton, 2004; Stø, Strandbakken, Throne-Holst, & Vittersø, 2004).

A growing recognition of the need to develop a more context-conscious approach to consumption has laid the basis for a body of literature that centres around the opposite end of the spectrum of theoretical perspectives on consumer behaviour. Purporting to correct the imbalances underlying the choice-based models of consumption, it links consumer subjectivities to particular social contexts and draws attention to a wide range of social relations, interactions, and processes in which consumer practices are contained. A distinctive body of research conducted within the framework of Consumer Culture Theory (Arnould & Thompson, 2005) emphasises and explores the multiple ways in which consumer practices and identities are shaped by a wide range of socio-cultural forces and embedded within the broader political, economic, and marketplace contexts. Various authors have taken consumer research to the social level by linking consumer ideas and actions to different levels of structure (Dobscha & Ozanne 2001; Oswald, 1999; Wallendorf, 2001); exploring the forging of shared beliefs, meanings, practices, and collective consumer identifications ensuing from them (Holt, 1997; Kozinets, 2002; Muniz & O'Guinn, 2001); and analysing the role of the marketplace as a source of symbolic materials for the production of individual and communal consumer identities (Askegaard & Kjeldgaard, 2002; Coskuner-Balli & Thompson, 2009; Coulter et al., 2003; Gopaldas, 2014).

The practice-based approach has arguably been the most influential among the theoretical developments targeting the social roots of consumption activities. Within it, consumption is understood to be embedded in everyday practices, routines, and relationships, centred around achieving some other targets – consumption, therefore, is not the end goal and has no intrinsic value, but occurs *within* and *for the sake of* other activities (Warde, 2005). Consequently, consumer choices are conceived of as functional elements in social practices rather than as expressions of an individual's wants, desires, and needs: "the logic of consumption is found not in the selection of items but in the practices within which they are utilized" (Warde, 2015, p. 118). Accordingly, the individual consumer is no longer seen as the unit of analysis that matters most; instead, the focus of scientific attention and empirical efforts shifts towards practices, their social constitutions and contexts. As Wheeler (2012, p. 91) points out, within the practice paradigm "interest moves away from attitudes and behaviours of an active consumer and instead concentrates on the 'do-ability' of practical performances and how these negotiated and shaped by social and institutional contexts".

Practice theories have quickly caught the wave of contemporary sociological thinking and become a large player in the field of consumer research. Since the beginning of the 21st century, the practice-based approach has been informing empirical work on sustainable consumption, drawing attention to the use of environmentally problematic commodities such as energy and water in the course of reproduction of mundane, taken-for-granted, symbolically inconspicuous practices and routines (e.g. Evans, 2011; Shove, 2003). Two major practice-theoretical programmes for sustainable consumption, as identified by Welch and Warde (2015), are those developed by Gert Spaargaren (2011) and Elisabeth Shove (2003). Whilst the two approaches have sprung from the same theoretical ground, differences can be discerned in their positioning of consuming agents vis-à-vis structural and systemic forces. Spaargaren situates individual consumers within social structures through the concept of environmental power, which refers to the capacity of citizen-consumers to reduce the environmental impact of consumption/production practices controlled by other social actors.

Shove (2003), however, goes as far as to completely remove individual meanings and actions from the research agenda for sustainable consumption and focuses on the relation between institutions, infrastructures, and technologies on the one hand and social conventions, understandings, and practices on the other. Shove's approach, developed in her landmark book *Comfort, Cleanliness and Convenience: The Social Organization of Normality* (2003) and subsequent publications (e.g. Shove, 2010; Shove, Pantzar, & Watson, 2012), promises to provide an understanding of the nature of social change required to achieve desired behavioural shifts in the sphere of consumption and sustainability (Shove et al., 2012). The explanatory value of practice theory, Shove argues, lies in its potential to provide insight into the ways in which people are recruited into practices and illuminate the dynamics of emergence, reproduction, and transformation of social practices in the course of daily life.

As Welch and Warde (2015) note, different versions of practice theory are united by the intent to "undermine the traditional individual-nonindividual divide by availing themselves of features of both sides" (Schatzki, 2001, p. 14). However, this purportedly anti-dualistic perspective creates more ontological problems than it solves. By refusing to draw a distinction between agential and structural properties, practice theories fall prey to the fallacy of "central conflation" (Archer, 2007), which relates structure and agency at the expense of their ontological and analytical integrity, and hence precludes understanding of how and with what consequences their interaction occurs. In its stronger version, practice-based perspective slips into "downward conflation": here, it presupposes an ontology in which practices are seen as the source of both social order, for they are "not merely 'sites' of interaction but are, instead, ordering and orchestrating entities in their own right" (Shove & Walker, 2010, p. 471), and individuality, since "It is practices that 'produce' and co-constitute individuals ... not the other way round" (Spaargaren, 2013, p. 233). Such a view of reality conflicts with a relational, inter-dependent, and inter-subjective social ontology, the relational character of which implies that structure cannot override agency (Donati, 2010), and the inter-subjectivity of which gives rise to ethical and moral intentions and actions, which themselves play an important part in structuring our societies (Berman, 2002).

The assumption of the ontological inferiority of individuals reverberates through the accounts of authors attempting to understand ethical consumption by dispensing with the image of the sovereign consumer and the idea of ethical shopping as a consumer-driven phenomenon. Jacobsen and Dulsrud (2007) explicitly question the existence of the "active consumer" and call for a more realistic understanding of the role of consuming agents in shaping the modern economy. They provide a detailed exposé of a wide range of strategically oriented actors, responsible for creating and shaping the ethical consumption phenomenon. Among them are NGOs, charities, and campaign groups advancing their own agenda, there is the corporate sector in search of profitable markets, and governments eager to lay the burden of responsibility for addressing environmental and societal challenges on the shoulders of citizens. These actors, whose divergent interests converge on a common goal of creating and governing the ethical consumer, encourage and enable the enactment of ethical consumer subjectivities through strategically conceived tools and techniques, such as labelling, campaigning, surveys, and polls. Claiming that "the consumer role is plastic and open for business interests, civic society organizations, and governmental agencies to mold" (Jacobsen & Dulsrud, 2007, p. 473), the authors leave little room for consumer agency and more complex dialectical relations between individual and social forces.

Barnett and colleagues (2010) also locate the drivers of ethical consumption in wider social, political, and market systems, which cultivate individuals as ethically minded consumers, acting in line with the principles of sustainability, ecological well-being, and respect for human rights. Thinking about ethical consumption in terms of power relations and drawing on Foucault's concept of governmentality, they bring into view the various agents who, while not ordinarily thought of as consumers, play a key role in the politicisation of consumption and recruitment of ordinary consumers "into broader projects of social change" (Barnett et al., 2005, p. 23). For Barnett et al. (2005, p. 23), ethical consumption involves both "a governing of consumption", which refers to attempts by collective actors to motivate and make possible certain types of behaviours and practices, and "a governing of the consuming self" – the process of cultivating one's own subjectivity through self-consciously responsible choices. The governmentality theme is elaborated upon in Clarke et al.'s (2007) work on the politics of ethical consumerism, which draws attention to the diversity of agents, strategies, and technologies responsible for the production of an ethical citizen-consumer. Here too, ethical consumption is conceptualised as ways in which an array of strategically motivated actors, including the state, corporations, and non-governmental organisations, deliberately and systematically create opportunities for ethically minded individuals to express their moral dispositions through marketplace actions. The mobilisation of individuals as ethical consumers, Clarke et al. argue, is achieved through the use of particular strategies, technologies, and devices which enable people to act in ethical ways when presented with options and allow the figure of an ethical consumer to be publicly visible in order to further the ethical consumption agenda. For example, consumer surveys and polls are used to generate data, such as statistics tracking the growth of the ethical goods and services sector, that can be used for raising public awareness, exerting pressure on industry players, and obtaining support for policy change.

To take another example from the repertoire of "narrative and practical resources" (Clarke et al., 2007, p. 235) involved in the making of a conscientious consumer, ethical labelling facilitates ethical consumer conduct by enabling individuals to make distinctions between ethical and conventional products. It is noteworthy that the role of ethical labelling in the working up of responsible consumers has also been explained in a way presupposing a much more active and reflexive agent than those engaged in a system-level analysis of ethical consumerism are willing to accommodate in their theorisations. Contrary to the representation of ethical labels as merely a practical tool "for turning *oughts* into *cans*" (Barnett et al., 2005, p. 31), many commentators argue for the ideological role of ethical labelling and its potential to contribute to consumers' moral conversion. According to Goodman (2004), ethical labels and promotion materials act as "translation devices", pulling individuals in the direction of more ethical food choices. Goodman and Goodman's (2001, p. 111) work on the geographies of sustainable consumption offers a rich and nuanced discussion of the ways in which morally charged texts and images generate discourses and narratives that enable fair-trade networks to "'lengthen' across the spaces of consumption, to work against and translate actors from more conventional agrofood networks". Bildtgård (2008) provides further testimony to the ethics-inducing potential of ethical labels, which he construes as time- and space-transcending devices that revive consumers' sense of responsibility by bridging the physical and cognitive gap between consumers and producers. These interpretations presuppose the existence of active, interpretive, meaning-making consumers who reflexively engage with the rhetoric and imagery of ethical labelling and allow them to interact with their subjective moral dispositions and beliefs, as opposed to using labels merely as a means to confirm earlier decisions made under the influence of objective social forces. It is, therefore, unsurprising that such perspectives on ethical labelling are not favoured by those who prefer to steer away from the image of the active consumer reflexively negotiating identity transitions through marketplace meaning-making. Clarke et al. (2007, p. 231) explicitly argue that

> the discursive interventions used in ethical consumption campaigns … are not primarily aimed at encouraging generic consumers to recognise themselves for the first time as 'ethical' consumers. Rather, they aim to provide information to people already disposed to support or sympathise with certain causes; information that enables them to extend their concerns and commitments into everyday consumption practices.

The assumption that individual subjectivity plays a negligible role in driving ethical consumerism is further reinforced by the claim that "it is acts, not identities or beliefs, which matter in mobilising the presence of 'ethical consumers' in the public realm" (Clarke et al., 2007, p. 241). Clarke et al. amplify the negation of the active consumer model in discussing local shopping as a practice in which "the exercise of 'choice' is shaped by systems of collective provisioning over which consumers have little direct influence" (2007, p. 239). The view of ethical consumption as a structural rather than agent-driven phenomenon is supported by Wheeler (2012), whose enquiry into the politics of the fair-trade movement underscores the key role of the systems of collective provision in mobilising and regulating consumer engagement with fair-trade goods.

On the whole, socio-centric perspectives are clearly juxtaposed against explanations of consumer behaviour in terms of the individual actor. In a battle against the "orthodoxy of the 'active consumer' in the social sciences" (Trentmann, 2006, p. 3), their proponents erase the image of an ethical consumer as an agent of active choice and ethical practices as expressions of individual liberty of conscience and thought. Practice theorists, for example, not only present social practices as "the principal steering device of consumption" (Warde, 2005, p. 145), but they also consider them to be "the primary source of desire, knowledge and judgment" (ibid.). Among studies conveying this view is Hards' (2011) enquiry into the process of development of personal environmental values. Hards' analysis is grounded in the practice-based conception of values as inherent components of social practices which form outside rather than within persons, namely, through encounters with ideas circulating and resonating in the wider society. These broadly shared ideas and understandings determine not only what people come to believe in, but also how they choose to enact their beliefs: agents perform practices, it is argued, in the ways that conform to commonly accepted standards and norms.

This practice-based outlook, which both underpins and is further backed up by Hards' account of nature-related values, can be challenged on several grounds. First, Hards' analysis locates the roots of participants' environmental beliefs in a wide range of apparently disparate life experiences, including rather trivial encounters with animals and much more extravagant incidents such as the use of psychedelic mushrooms. The obviously arbitrary nature of belief-inducing experiences described in the study runs contrary to practice theorists' view that values are neither subjectively developed nor personally possessed but are mere reflections of socially dominant ideas and norms (were the latter state of affairs true, we would be observing much higher degrees of conformity than our societies can currently boast). The argument becomes even less tenable in light of the fact that environmentalism harbours multiple, oftentimes inconsistent, and even contradictory practices and beliefs. The anchoring of individual pro-environmental behaviours in "a broadly shared conception of what it means to live a low-carbon life" (Hards, 2011, p. 26) is undermined by the ambiguity surrounding the environmental impacts of various human practices and activities. As Cherrier (2007, p. 322) argues, "there cannot be, for example, a regime of truth about recycling when scientists disagree on the evidence, country representatives disagree on the outcomes, and commentators' opinions change continuously". Likewise, the messages about how to estimate and manage one's carbon footprint circulating within the scientific, media, and general-public discourses are far from univocal. For instance, organic foods are commonly understood to have lower CO_2 footprint than conventionally grown produce; at the same time, warnings abound that the environmental benefits of organic products shipped over lengthy distances are, to say the least, questionable and most likely to be severely compromised if not altogether outweighed by the negative impact of transport emissions. Given the lack of consensus about environmental evils and goods, the divergence in individual understandings and performances of environmentalism seems unavoidable, as noted by Cherrier (2007, p. 322):

> The pluralization of expert systems and greater access to information prompts multiple and often contradictory opinions about the "what" and "how" of ethical consumption (Beck, 1999) such that what seems good or ethical for one may not be so for another.

Faced with a wide range of ethical problems, and an even wider range of opinions on how to address any given one, individuals will have to exercise their own moral judgment and become the lead authors in the making of their ethical selves and in defining a set of practices that best corresponds to their self-concepts. This subjective identity and lifestyle work – the privilege and burden of a postmodern agent – is what gives rise to distinctive value positions and idiosyncratic performances of environmental practices by agents whose courses of actions are steered by their unique patterns of concerns. Thus, consumers may decide to go local or, conversely, support farming families in the developing world; they may try to "grow their own" in a bid to opt out of the corporate food industry, or vote with their supermarket trolleys to drive change in production practices; they may choose to invest in organic or fair trade depending on whether a healthy environment or social justice feels like the most urgent priority.

Given that there are many possible ways of being an ethical consumer, individual perceptions and enactments of consumption ethics cannot be mere reflections of socially prescribed understandings and actions (although, of course, the latter provide reference points against which alternatives can be judged). It seems indeed only natural that "in a pluralistic and complex world, things that seem ethical to one person may not mirror the general stance on an issue" (Cherrier, 2007, p. 331). Empirical evidence supports this conjecture: based on a study of individual consumption of fair trade, Adams and Raisborough (2010) conclude that individual ethics may not always straightforwardly correspond to normative frameworks and that a wide range of positions are available for consumers to occupy in response to ethical demands. Hards' own findings seem to lend support to the argument: as the aforementioned study reports, some of the environmentalists Hards interviewed engaged in climate-change activism to promote social justice rather than nature-related values typically associated with pro-environmental behaviours. This testifies to the subjects' ability to shape their practices in accordance with their personal concerns and their subjective understandings of how it would be best to act upon them. What seems to matter here is the perceived congruence between subjects' chosen lifestyles and their ultimate concerns – what they care about and what they want to achieve in light of a deep sense of commitment – as opposed to how well their practices conform to broadly accepted definitions and standards of ethical consumer conduct. I return to this point in Chapter 7, which provides empirical evidence demonstrating the role of consuming agents in shaping and moulding the ethical consumption phenomenon.

In denying individuals the ability to actively define and own their values and practices, the practice-based framework essentially takes aims at another fundamental aspect of human make-up, namely, the capacity for reflexivity. Indeed, agential reflexivity has been largely displaced from the socio-centric accounts of consumption. Individual practices being in the custody of social forces, there remains little room or, indeed, need for consuming agents to exercise their potential for reflexive deliberation. Naturally then, practice theorists have little regard for the reflexive consumer, and the view that "consumption occurs often entirely without mind" (Warde, 2005, p. 150) prevails among the proponents of the practice-based framework. Some, however, have taken a more moderate stance on consumer reflexivity. Wheeler, for example, appears somewhat less eager to expel reflexivity from sociological accounts of consumer activities. In her view, the practice-based perspective accommodates both routine and reflexive consumption and allows enough room for human agency even in the context of a concerted effort to regulate consumer behaviour. Wheeler suggests that the increased ascription of societal responsibility to consumers "can create an occasion" for individuals to reflexively evaluate and modify their shopping habits (Wheeler, 2012, p. 91). Following Warde, she points out that practices "are internally differentiated on many dimensions" (Warde, 2005, p. 138 cited in Wheeler, 2012, p. 89) and that their enactments are conditional upon "time, space, and social context" (Warde, 2005, p. 139 cited in Wheeler, 2012, p. 90).

This, however, is a rather feeble defence of agential reflexivity. First, what is left of humans' ability to actively draw on their reflexive powers if those are only evoked when social conditions "create an occasion" and command people to do so? Further, what possibilities are really left for individual subjectivity to contribute to the formation and transformation of practices if their meanings and forms are restricted by social organisation and their internal variations arise solely due to the differing outer contexts? This seems only to reinforce practice theorists' over-socialised view of consumption as devoid of agency, intent, and volition and their dismissal of reflexivity as a steering force of consumer conduct. While socio-centric perspectives go a long way towards illuminating structural influences shaping consumption, their explanatory potential is hampered by the dismissal of the figure of an active, ethically conscious consumer. Practice-theorists' interpretation of consumption as a "relatively unconscious form of life" (Soper, 2009, p. 12) is evidently inadequate for explaining ethical consumption activities. Their proposed framework, in which "the concept of 'the consumer' ... evaporates" (Warde, 2005, p. 146), does not allow for the exploring of the phenomenon of ethical consumption at the individual level and passes over a range of micro-level factors that are critical for understanding consumer behaviour. Likewise, the view of ethical choices as little more than socially orchestrated enactments of externally imposed and strategically governed roles and identities, as favoured by Clarke and colleagues, precludes important insights into the world of subjective meanings surrounding ethical consumer decisions.

As this discussion has shown, existing research on ethical consumption is tied up with the dominant theoretical frameworks in which the consumer is presented as either an agent of free choice or a passive bearer of socially defined and regulated practices. Reproducing the core ontological and analytical presuppositions of these theoretical outlooks, the vast majority of studies on ethical consumption have taken uncompromising approaches where either consumer agency or social structure was emphasised and subjected to scrutiny. The key reason for the ineffectiveness of this body of work in developing a comprehensive and nuanced account of ethical consumer behaviour, individual and collective, has been the tendency to leave out of sight the full spectrum of forces at work in creating and steering consumption. Each strand of research has only been able to achieve a partial understanding of the phenomenon whilst leaving many important aspects and dimensions of ethical consumer practices and experiences unacknowledged and unexplored. On the one hand, agency-focused perspectives have attained noticeable progress in offering an enhanced understanding of the subjective motives and meanings attached to ethical consumer choices, but neither an adequate account nor even an explicit acknowledgement of the contexts in which these choices are made and the external factors that determine them has ensued. On the other hand, socio-centric approaches have encouraged recognition of the social underpinnings of consumer behaviour and the embeddedness of individual choices in the social and material organisation of life, while appearing to neglect the ways in which consumer agency and individual subjectivity interact with and respond to the social order. They, therefore, fall short of effectively accounting for aspects of individual engagement with ethical consumption and adequately explaining the variations in its understandings and performances among the consuming agents.

Thus, applied in isolation, neither agency-focused frameworks nor socio-centric paradigms proved sufficient for enabling consumer studies to provide a much-needed understanding of the complex ensemble of individual and structural factors that determine consumption phenomena. Empirical research informed by either of these two perspectives inevitably leads to a one-dimensional view of consumer behaviour: it either reduces its social aspect to an aggregate of individual actions or dissolves the consuming agent in society, attributing his decisions solely to structural imperatives and systemic prescriptions. Soper (2007, p. 217) diagnoses this unhelpful duality:

> Consumer behaviour that is treated as a matter of existential choice, through the one optic, is viewed through the other, as the altogether less voluntary effect of transcendent economic and social structures and their systemic pressures and forms of social governance.

The recognition of the need to surpass the apparent limitations of one-sided approaches to consumption has been growing among social theorists over the past years. Sassatelli (2007, p. 107) has urged consumer studies to "overcome that moralistic swing of the pendulum which … either celebrates consumption as a free and liberating act, or denigrates it as a dominated and subjugated act". Likewise, Halkier (2010, p. 14) recommends "the complexity position", which acknowledges the everyday complexities of consumption and "seeks to unfold both agency capacities and the social conditioning of ordinary consumers". Johnston (2007, p. 233) specifically presses for a dialectic approach to ethical consumption that "helps us avoid naïve optimism, or determinist pessimistic accounts of consumer-focused projects for social justice and sustainability" and that "recognizes that meaning and agency are present in consumption decisions but takes seriously the structural conditions shaping consumer agency". Yet, albeit the willingness to move away from the simplicity of one-sided views on consumption and develop a more sophisticated, multi-dimensional understanding of the phenomenon is apparently growing, reframing consumption along the suggested lines is far from a *fait accompli*. First, while quite a few authors have theorised an integrated perspective, little, if any, academic effort has gone into putting the theory into practice, and the lack of empirical research exploring how both agency and structure manifest themselves in consumption persists. It is also problematic that many commentators continue to place hope in the theories of practice to steer research towards a balanced approach to consumer behaviour (e.g. Halkier, 2010; Jacobsen & Dulsrud, 2007; Sassatelli, 2007; Spaargaren, 2011). Illustrative is Warde's (2015, p. 129) conclusion, derived at the end of a comprehensive review of more than four decades of sociological research on consumption:

> From a sociological point of view, it is much better to unseat the dominant model of the sovereign consumer and replace it with a conception of the

socially conditioned actor, a social self, embedded in normative and institutional contexts and considered a bearer of practices.

Yet, it seems to me that merely replacing the short-sightedness of agency-focused perspectives with the partiality of practice-based approaches is not the way forward for consumer research if a long-sought recognition of the nuanced complexity of consumption is to be achieved. While Jacobsen and Dulsrud's (2007, p. 469) appeal to reject the belief in the active consumer as "a universal entity, available across nations and time" is clearly justified, this should not lead the field to dispense with the concept of agency altogether, or completely deny consumers the liberty of thought, conscience, and choice, or reduce individual acts and decisions to involuntary effects of systemic pressures. I argue that the questions of structure and agency should remain on the agenda of consumer research, for the biases inherent in the current understanding of consumer behaviour can only be redressed by rethinking the act of consumption as one where a complex interweaving of agential and structural powers occurs. Ultimately, what needs to be acknowledged is that "practices of consumption are meaningful for people even if they are not entirely free or always consequential; they are enclosed in mechanisms of power even if these are not deterministic" (Sassatelli, 2007, p. 107). Through the application of a critical realist approach, this book charts a path towards understanding the dialectical relationship between consuming agents and the external world.

References

Adams, M. (2003). The reflexive self and culture: a critique. *British Journal of Sociology, 54*(2), 221–238.

Adams, M., & Raisborough, J. (2008). What can sociology say about FairTrade?: class, reflexivity and ethical consumption. *Sociology, 42*(6), 1165–1182.

Adams, M., & Raisborough, J. (2010). Making a difference: ethical consumption and the everyday. *The British Journal of Sociology, 61*(2), 256–274.

Allison, G. (2009). Pursuing status through ethical consumption? In Tojib, D. (Ed.), *Sustainable Management and Marketing*. ANZMAC Conference, Monash University, Melbourne, 30 November – 2 December.

Archer, M. (2000). *Being Human: The Problem of Agency*. Cambridge, UK: Cambridge University Press.

Archer, M. (2007). *Making Our Way through the World*. Cambridge, UK: Cambridge University Press.

Arnould, E. J., & Thompson, C. J. (2005). Consumer culture theory (CCT): twenty years of research. *Journal of Consumer Research, 31*(4), 868–882.

Arvola, A., Vassallo, M., Dean, M., Lampila, P., Saba, A., Lähteenmäki, L., & Shepherd, R. (2008). Predicting intentions to purchase organic food: the role of affective and moral attitudes in the Theory of Planned Behaviour. *Appetite, 50*(2–3), 443–454.

Askegaard, S., & Kjeldgaard, D. (2002). The water fish swim in? Relations between culture and marketing in the age of globalization. In Knudsen, T., Askegaard, S., & Jørgensen, N. (Eds.), *Perspectives on Marketing Relationships* (pp. 13–35). Copenhagen: Thomson.

Askegaard, S., & Linnet, J. (2011). Towards an epistemology of consumer culture theory: phenomenology and the context of context. *Marketing Theory, 11*(4), 381–404.

Barnett, C., Cloke, P., Clarke, N., & Malpass, A. (2005). Consuming ethics: articulating the subjects and spaces of ethical consumption. *Antipode, 37*(1), 23–45.

Barnett, C., Cloke, P., Clarke, N., & Malpass, A. (2010). *Globalizing Responsibility: The Political Rationalities of Ethical Consumption*. West Sussex, UK: John Wiley & Sons.

Becker, G. S. (1996). *Accounting for Tastes*. Cambridge, MA: Harvard University Press.

Berman, M. (2002). Merleau-Ponty and Nagarjuna: enlightenment, ethics and politics. *Journal of Indian Philosophy and Religion, 7*(Oct), 99–129.

BildtgÅrd, T. (2008). Trust in food in modern and late-modern societies. *Social Science Information, 47*(1), 99–128.

Brennan, G., & Tullock, G. (1982). An economic theory of military tactics: methodological individualism at war. *Journal of Economic Behavior and Organization, 3*(3), 225–242.

Chang, H. (2014). *Economics: The User's Guide*. New York, NY: Bloomsbury Press.

Cherrier, H. (2006). Consumer identity and moral obligations in non-plastic bag consumption: a dialectical perspective. *International Journal of Consumer Studies, 30*(5), 515–523.

Cherrier, H. (2007). Ethical consumption practices: co-production of self-expression and social recognition. *Journal of Consumer Behaviour, 6*(5), 321–335.

Clarke, N., Barnett, C., Cloke, P., & Malpass, A. (2007). Globalising the consumer: doing politics in an ethical register. *Political Geography, 26*(3), 231–249.

Coskuner-Balli, G., & Thompson, C. (2009). Legitimatizing an emergent social identity through marketplace performances. In McGill, A. L. & Shavitt, S. (Eds.), *Advances in Consumer Research* (Vol. 36, pp. 135–138). Duluth, MN: Association for Consumer Research.

Coulter, R., Price, L., & Feick, L. (2003). Rethinking the origins of involvement and brand commitment: insights from postsocialist central Europe. *Journal of Consumer Research, 30*(2), 151–169.

De Groot, J., & Steg, L. (2009). Mean or green? Values, morality and environmental significant behavior. *Conservation Letters, 2*(2), 61–66.

Diner, H. (2001). *Hungering for America*. Cambridge, MA: Harvard University Press.

Dobscha, S., & Ozanne, J. (2001). An ecofeminist analysis of environmentally sensitive women using qualitative methodology: the emancipatory potential of an ecological life. *Journal of Public Policy and Marketing, 20*(2), 201–214.

Donati, P. (2010). *Relational Sociology: A New Paradigm for the Social Sciences*. London, UK: Routledge.

Donati, P., & Archer, M. S. (2015). *The Relational Subject*. Cambridge, UK: Cambridge University Press.

Evans, D. (2011). Beyond the throwaway society: ordinary domestic practice and a sociological approach to household food waste. *Sociology, 46*(1), 41–56.

Fischler, C. (1980). Food habits, social change and the nature/culture dilemma. *Social Science Information, 19*(6), 937–953.

Fischler, C. (1988). Food, self and identity. *Social Science Information, 27*(2), 275–292.

Flam, H. (2000). *The Emotional Man and the Problem of Collective Action*. Frankfurt am Main, Germany: Peter Lang.

Fraj, E., & Martinez, E. (2007). Ecological consumer behaviour: an empirical analysis. *International Journal of Consumer Studies, 31*(1), 26–33.

Gabriel, Y., & Lang, T. (2006). *The Unmanageable Consumer*. Thousand Oaks, CA: Sage Publications.

Giddens, A. (1991). *Modernity and Self–identity*. Stanford, CA: Stanford University Press.

Goodman, D., & Goodman, M. (2001). Sustaining foods: organic consumption and the socio-ecological imaginary. In Cohen, M. & Murphy, J. (Eds.), *Exploring Sustainable Consumption: Environmental Policy and the Social Sciences* (pp. 97–119). Oxford, UK: Elsevier.

Goodman, M. (2004). Reading fair trade: political ecological imaginary and the moral economy of fair trade foods. *Political Geography*, *23*(7), 891–915.

Gopaldas, A. (2014). Marketplace sentiments. *Journal of Consumer Research*, *41*(4), 995–1014.

Guthman, J. (2002). Commodified meanings, meaningful commodities: re-thinking production-consumption links through the organic system of provision. *Sociologia Ruralis*, *42*(4), 295–311.

Halkier, B. (2010). *Consumption Challenged: Food in Medialised Everyday Lives*. Farnham, UK: Ashgate Publishing Group.

Hards, S. (2011). Social practice and the evolution of personal environmental values. *Environmental Values*, *20*(1), 23–42.

Hilton, M. (2004). The legacy of luxury: moralities of consumption since the 18th century. *Journal of Consumer Culture*, *4*(1), 101–123.

Holt, D. (1997). Poststructuralist lifestyle analysis: conceptualizing the social patterning of consumption. *Journal of Consumer Research*, *23*(4), 326–350.

Jacobsen, E., & Dulsrud, A. (2007). Will consumers save the world? The framing of political consumerism. *Journal of Agricultural and Environmental Ethics*, *20*(5), 469–482.

John, A., Klein, J., & Smith, N. (2002). Exploring motivations for participation in a consumer boycott. *Advances in Consumer Research*, *29*(1), 363–369.

Johnston, J. (2007). The citizen-consumer hybrid: ideological tensions and the case of Whole Foods Market. *Theory and Society*, *37*(3), 229–270.

Kozinets, R. (2002). Can consumers escape the market? Emancipatory illuminations from Burning Man. *Journal of Consumer Research*, *29*(1), 20–38.

Kozinets, R., & Handelman, J. (1998). Ensouling consumption: a netnographic exploration of the meaning of boycotting behavior. *Advances in Consumer Research*, *25*(1), 475–480.

Lang, T., & Heasman, M. (2004). *Food Wars*. London, UK: Earthscan.

Lang, T., Barling, D., & Caraher, M. (2009). *Food Policy*. Oxford, UK: Oxford University Press.

Lekakis, E. (2013). *Coffee Activism and the Politics of Fair Trade and Ethical Consumption in the Global North*. Basingstoke, UK: Palgrave Macmillan.

McKie, L., & Wood, R. (1992). People's sources of recipes: some implications for an understanding of food-related behaviour. *British Food Journal*, *94*(2), 12–17.

Mennell, S., Murcott, A., & van Otterloo, A. (1992). The sociology of food: eating, diet and culture. *Current Sociology*, *40*(2), 1–125.

Micheletti, M. (2003). *Political Virtue and Shopping*. New York, NY: Palgrave Macmillan.

Miller, D. (1998). *A Theory of Shopping*. Cambridge, UK: Polity Press.

Miller, M. B. (1981). *The Bon Marché: Bourgeois Culture and the Department Store, 1869–1920*. Princeton, NJ: Princeton University Press.

Moisander, J., & Pesonen, S. (2002). Narratives of sustainable ways of living: constructing the self and the other as a green consumer. *Management Decision*, *40*(4), 329–342.

Muniz, A. M., & O'Guinn, T. C. (2001). Brand community. *Journal of Consumer Research*, *27*(4), 412–432.

Murcott, A. (1983). *The Sociology of Food and Eating*. Aldershot, UK: Gower.

Murdoch, J., & Miele, M. (1999). "Back to nature": changing "worlds of production" in the food sector. *Sociologia Ruralis, 39*(4), 465–483.

Newholm, T. (2005). Case studying ethical consumers' projects and strategies. In Harrison, R., Newholm, T., & Shaw, D. (Eds.), *The Ethical Consumer* (pp. 107–124). London, UK: Sage Publications.

Oswald, L. (1999). Culture swapping: consumption and the ethnogenesis of middle-class Haitian immigrants. *Journal of Consumer Research, 25*(4), 303–318.

Sassatelli, R. (2007). *Consumer Culture: History, Theory and Politics.* Los Angeles, CA: Sage Publications.

Sassatelli, R. (2012). Consumer identities. In A. Elliott (Ed.), *Routledge Handbook of Identity Studies* (pp. 236–253). London, UK: Routledge.

Sayer, A. (2011). *Why Things Matter to People.* Cambridge, UK: Cambridge University Press.

Schatzki, T. (2001). Introduction: practice theory. In Schatzki, T., Knorr–Cetina, K. & von Savigny, E. (Eds.), *The Practice Turn in Contemporary Theory* (pp. 1–14). London, UK: Routledge.

Shaw, D. (2007). Consumer voters in imagined communities. *International Journal of Sociology and Social Policy, 27*(3/4), 135–150.

Shaw, D., & Shiu, E. (2003). Ethics in consumer choice: a multivariate modelling approach. *European Journal of Marketing, 37*(10), 1485–1498.

Shove, E. (2003). *Comfort, Cleanliness and Convenience.* Oxford, UK: Berg.

Shove, E., & Walker, G. (2010). Governing transitions in the sustainability of everyday life. *Research Policy, 39*(4), 471–476.

Shove, E., Pantzar, M., & Watson, M. (2012). *The Dynamics of Social Practice.* Los Angeles, CA: Sage Publications.

Soper, K. (2007). Re-thinking the "good life": the citizenship dimension of consumer disaffection with consumerism. *Journal of Consumer Culture, 7*(2), 205–229.

Soper, K. (2008). Alternative hedonism, cultural theory and the role of aesthetic revisioning. *Cultural Studies, 22*(5), 567–587.

Soper, K. (2009). Introduction: the mainstream of counter-consumerist concern. In Soper, K., Ryle, M. & Thomas, L. (Eds.), *The Politics and Pleasures of Consuming Differently* (pp. 1–21). New York, NY: Palgrave Macmillan.

Spaargaren, G. (2011). Theories of practices: agency, technology, and culture: exploring the relevance of practice theories for the governance of sustainable consumption practices in the new world-order. *Global Environmental Change, 21*(3), 813–822.

Starr, M. (2009). The social economics of ethical consumption: theoretical considerations and empirical evidence. *The Journal of Socio-Economics, 38*(6), 916–925.

Starr, P. (2011). Ethical consumption, sustainable production and wine. In Lewis, T., & Potter, E. (Eds.), *Ethical Consumption: A Critical Introduction* (pp. 131–140). London, UK: Routledge.

Stø, E., Strandbakken, P., Throne-Holst, H., & Vittersø, G. (2004). *Potentials and limitations of environmental information to individual consumers.* Paper presented at the 9th European Roundtable on Sustainable Consumption and Production, Bilbao, Spain, May.

Tallontire, A., Rentsendorj, E., & Blowfield, M. (2001). *Ethical Consumers and Ethical Trade.* Kent, UK: Natural Resources Institute.

Torjusen, H., Lieblein, G., Wandel, M., & Francis, C. (2001). Food system orientation and quality perception among consumers and producers of organic food in Hedmark County, Norway. *Food Quality and Preference, 12*(3), 207–216.

Trentmann, F. (2006). *The Making of the Consumer.* Oxford, UK: Berg.

Tucker, K. (1998). *Anthony Giddens and Modern Social Theory*. London, UK: Sage Publications.

Veblen, T. (2009). *The Theory of the Leisure Class*. Oxford, UK: Oxford University Press.

Wallendorf, M. (2001). Literally literacy. *Journal of Consumer Research*, *27*(4), 505–511.

Warde, A. (1997). *Consumption, Food, and Taste*. London, UK: Sage Publications.

Warde, A. (2005). Consumption and theories of practice. *Journal of Consumer Culture*, *5*(2), 131–153.

Warde, A. (2015). The sociology of consumption: its recent development. *Annual Review of Sociology*, *41*(1), 117–134.

Wattanasuwan, K. (2005). The self and symbolic consumption. *Journal of American Academy of Business*, *6*(1), 179–184.

Welch, D., & Warde, A. (2015). Theories of practice and sustainable consumption. In L. Reisch, & J. Thøgersen (Eds.), *Handbook of Research on Sustainable Consumption* (pp. 84–100). Cheltenham, UK: Edward Elgar.

Wheeler, K. (2012). *Fair Trade and the Citizen-Consumer*. Basingstoke, UK: Palgrave Macmillan.

Wilk, R. (2001). Consuming morality. *Journal of Consumer Culture*, *1*(2), 245–260.

2 Ethical consumption and critical realism

My enquiry into the inner and outer lives of ethical food consumers is underpinned by critical realism and draws specifically on a critical realist account of identity proposed by Margaret Archer (2000, 2007). At the heart of this account is the concept of reflexivity, defined as "the mental ability, shared by all normal people, to consider themselves in relation to their (social) contexts and vice versa" (Archer, 2007, p. 4). Such understanding of agential reflexivity has critical implications for how we conceptualise, study, and interpret consumer behaviour. Conceived of as a fundamental human capacity to deliberate upon themselves and their environments, reflexivity leads one to dispense with "the portfolio model" (Hindess, 1990) of the human subject wherein individual actions are guided by desires, preferences, and beliefs derived from a pre-given and supposedly stable portfolio – the ontology shared by agency-focused approaches, whether based upon the expressive identity-seeking or the rational preference-driven consumer. In addition, the figure of the reflexive consumer does not tolerate the assumption of the ontological supremacy of social meanings, competencies, and routines, considered by practice theorists to be the key determinants of individual understandings and actions. The idea of reflexivity as a mediating force between structure and agency provides a conceptual tool for explaining human conduct in a way that accounts both for the influence of social structures on agential practices and the ability of individual agents to interact with and respond to the social order. It is, therefore, a prerequisite to the success of my analytical project, aiming to shed light on the ways in which ethically concerned individuals negotiate the constraining and habilitating properties of their social contexts in a continuous effort to achieve and sustain their desired identities.

In the next section, I draw on extant philosophical and sociological knowledge to establish the ontological status of reflexivity as an essential human property and a fundamental feature of personhood.

Reflexivity: an emergent subjective power

The rise of the concept of reflexivity in sociological accounts of the human condition is linked to the unfolding of what has been termed a "post-traditional society" (Giddens, 1994) or "risk society" (Beck, 1992). A number of influential sociological thinkers have put forward the view that the demise of the class structure as an organising principle of social life has triggered an intensifying process of informalisation, reflected in a progressive decline of moral, aesthetic, and social standards of behaviour. Considered one of the most powerful societal trends of the second half of the 20th century (Wouters, 1986), it has thrown people into the state of profound social disembeddedness, manifest in the erosion of conventional ties of community, kinship, tradition, culture, and physical places, and loss of connectedness between individuals in social environments (Barrera, 1986). In the absence of collective prescriptions for social conduct, individual judgments have become key to attributing and maintaining social meaning. The following quote from Sörbom (as cited in Shaw, 2007, p. 143) succinctly describes the condition: "we have become released from collective and traditional authorities … it is now up to the individual to find whatever she or he perceives to be true – it can no longer be transferred from some higher power".

In the context of detraditionalisation, reflexivity has been singled out as a primary tool that allows highly individualised subjects to attempt to solve the problem of self-identity, that is to "produce, stage and cobble together their biographies themselves" (Beck, 1994, p. 13) in a society where one has "no choice but to choose how to be and how to act" (Giddens, 1994, p. 75). Another social factor claimed to be responsible for accelerating the transformation of individuals from passive into reflexive agents is the proliferation of new kinds of risks, created by the modern welfare society and left to increasingly self-dependent people to negotiate and deal with (Beck, 1992; Giddens, 1991).

In light of a new understanding of modern social conditions, the assumption that "people just get socialised 'in the mechanical way'" (Bradford, 2012) has become less and less tenable. The idea that human beings have the capacity and a natural tendency to be more or less reflexive as they make their way through the world is now broadly accepted, although many aspects of the phenomenon continue to fuel debates in contemporary research and theory in a number of disciplines. Archer argues strongly for the indispensability of reflexivity to the functioning of social subjects and the very existence of society as a whole. In her *Making our Way through the World*, she provides an extensive discussion and detailed defence of reflexivity as an essential human capacity and "a transcendentally necessary condition for the workings of any society" (Archer, 2007, p. 31). The requirement to be reflexive, as Archer indicates, is not weakened by our inevitable embeddedness in the social world, but essentially engendered by it. Societies rely for their existence and functioning on the ability of subjects to recognise themselves as parts of the society and as bearers of its norms – expectations, responsibilities, and duties attached to particular positions and roles. Socialisation, therefore, is an active process, which requires individuals to consciously consider themselves *vis-à-vis* their social environments, deliberate upon various positions available for them to occupy in relation to the social order (one may well end up being an anarchist), weigh them up in terms of attractiveness and associated costs, and, finally, internalise certain norms and reject certain others. In this way, Archer argues, the social world presupposes individual reflexivity, itself shaped by the social contexts in which it is exercised by agents.

An explicit acknowledgement of the social situatedness, limited potential, and inherent fallibility of reflexive thought is a distinct feature and a key strength of Archer's account of reflexivity. Whilst placing the capacity for reflexive reasoning at the very core of what it means to be a person, Archer consistently emphasises the limits of agential reflexivity by reminding us, throughout her works, that human knowledge is incomplete, specific, and partial for "subjects do not and cannot know everything that is going on" (2007, p. 23), that individual experiences represent only "that which is accessible to actors at any given time in its incompleteness and distortion and replete with its blind spots of ignorance" (1998, p. 369), and that "agents can only know themselves and their circumstances under their own descriptions, which are fallible, as is all our knowledge" (Archer, 2003, p. 15). Yet, for Archer, the fallibility of human reflexivity does not weaken its role in shaping agential conclusions about the self and its place in the world. The fact that our reflexive powers are inherently limited and error-prone cancels out neither our ability to engage in reflexive deliberations, nor our intrinsic propensity to do so. While, admittedly, subjects cannot and do not know everything, what they can do is reflect upon the world and what they care about in it, define what they are willing and able to commit to in life, and, in fulfilling their commitments – as best they know and as best they can – achieve their distinct personal identities. Crucially, Archer places our fallible – yet corrigible – reflexivity at the very heart of the identity processes, as we shall see in the next chapter.

An excursus to the philosophy of mind adds another layer of depth to our understanding of the relationship between reflexivity and identity. Elaborating on Locke, the philosopher Marya Schechtman (2005, p. 18) puts forward what she refers to as the "self-understanding" view of personal identity which foregrounds the importance of "being intelligible to ourselves" in the formation of a person's conception of the self as a continuous subject. On Schechtman's account, for a person to form the kind of self-conception that constitutes personal identity, "she must see her life as unfolding according to an intelligible trajectory, where present states follow meaningfully from past ones, and the future is anticipated to bear certain predictable relations to the present" (2005, p. 18). This kind of self-awareness involves a deliberate attempt on the part of the individual to make sense of her own experiences and their relation to her self-identity: to be a self-conscious person, Schechtman argues, is to take interest "in the character of our experience, and also in what we should do and what kind of person we should be" (2005, p. 18). What this implies is that "we are constantly self-monitoring, keeping track of how we are feeling, what we are doing, and what we are like" (ibid.), and it is this self-monitoring that gives individuals a sense of being a persistent and stable self throughout their lifespan. Pre-empting objections, Schechtman's notes that this continuous self-monitoring need not always be explicit or even obvious to ourselves: while in many instances we deliberately introspect and consciously examine the unfolding of things in our lives, most of this reflexive work occurs in the background. Further, she points out that the kind of self-perception, self-consciousness, and eventually self-understanding that, in her view, is vital for personal identity does not require that a subject has complete control over the course of her life or an absolute grasp of it, "only that she can see connections between how things were, how they are, and how they are likely to be" (Schechtman, 2005, p. 18). This important idea is neatly expressed in the following quote:

> Sometimes, however, scrutiny of our environment and of our conscious internal states still leaves us baffled about why we feel or act as we do. This unintelligibility threatens our integrity as self-conscious subjects—in the subject, as in the world more generally, there should be no events that are simply uncaused. This does not mean that we must fully understand all of our feelings or motives, but only that we should not be at a loss as to where to start in such self-understanding.

An excerpt from the novel *Headlong* (Frayn, 1999, pp. 126–127), which Schechtman refers to, illustrates such reflexive self-searching in an eloquent way:

> Odd, though, all these dealings of mine with myself. First I've agreed to a principle with myself, now I'm making out a case to myself and debating my own feelings and intentions with myself. Who is this self, this phantom internal partner, with whom I'm entering into all of these arrangements? (I ask myself.) Well, who am I talking to now? Who is the ghostly audience for the long tale I tell through every minute of the day? This silent judge sitting, face shrouded, in perpetual closed session?

In relaying the protagonist's reflexive musings with himself and over himself, Frayn essentially provides a non-academic account of internal conversation – a silent, inner self-talk in which people continuously engage as they are trying to make sense of their experiences and the world in which they live. The idea of self-monitoring as a basic part of the human condition is further elaborated in the work of the philosopher, Raymond Martin, who considers the nature of self and provides a phenomenological description of personal identity in his book, *Self-Concern* (1998). There he develops the notion of a perceiver-self as a means of understanding the feeling of being an outside observer of one's own body, actions, and thoughts – a mental experience which most of us will be familiar with. This admittedly illusory entity, the author argues, represents an essential feature of our psychological makeup which contributes fundamentally to one's awareness of the self as a stable and continuous subject. While the nature of the perceiver-self and the extent of its contributions to personal identity are far from determined, Martin's analysis of this ubiquitous mental phenomenon adds further weight to the claim that reflexive self-monitoring constitutes a pervasive element of individuals' inner lives.

It is hard not to notice parallels between philosophical views on self-monitoring and self-understanding and Archer's account of the nature and function of human reflexivity. Archer's approach to reflexivity is deeply philosophical in that she essentially ontologises reflexivity, that is, she construes reflexive thought not only as something that emerges when a person engages in a deliberate process of thinking about or contemplating certain things or events, but as the very essence of human nature. Further, there are apparent similarities between Archer's and Schechtman's thinking about self-conscious self-monitoring (reflexivity) and a person's sense of a coherent, continuous self (personal identity). Both Schechtman and Archer link our natural propensity for self-monitoring to our emergent capacity to deliberate upon and make sense of our feelings, thoughts, and behaviours, both attribute to reflexive self-monitoring, a key role in the formation of personal identity, and both emphasise the limits and fallibility of people's understanding of their internal states and external environments. Finally, akin to Schechtman and Martin, Archer places reflexive pursuits and achievements at the very core of her theorising about personhood.

It is based on these deep and extensive insights from the philosophical and sociological literature on reflexivity and identity that I take it to be ontologically true that "a subjective mental world of personal experiences exists" (Popper, 1972, p. 136), that all human beings possess "a generative ability for internal deliberation upon external reality" (Archer, 2003, p. 20), that this reflexive deliberation "leads to self-knowledge: about what to do, what to think and what to say" (Archer, 2003, p. 26) and constitutes a fundamental part of the identity processes. Consequently, I assign to the subjective power of reflexivity a leading part in the formation of ethical consumer identities: its task is to drive a continuous dialogue between one's inner self and the outer world and enable individuals to determine what truly matters to them, what they want to achieve in life, and how their identity-defining commitments can be fulfilled under the given conditions. Whilst defending the model of the reflexive agent, I guard against the assumption of context-transcendent, limitless reflexivity. The demand to recognise reflexivity as an intrinsic human capacity, underlying our sense of self as a distinct and continuous person, is not a demand to endorse the image of an all-seeing and all-knowing actor, capable of full discursive penetration of the subjective self and its objective conditions. Reflexivity that lays a claim to a complete self-understanding has been called into question by various thinkers reflecting on the construction of identities. In her influential article on the production of situated knowledges, Gillian Rose (1997) helpfully summarises the arguments against the possibility of total self-understanding by means of reflexive self-scrutiny. Elaborating upon Gibson-Graham's (1994, p. 206) problematisation of the idea of herself as "a centred and knowing subject who is present to myself and can be spoken for", Rose portrays a self as "un-centred, un-certain, not entirely present, not fully representable: this is not a self that can be revealed by a process of self-reflection". Presenting identity as relational, that is grounded in a sense of being different from the others, she asserts the impossibility of fully knowing one's "otherness" which, in turn, precludes the possibility of fully knowing oneself. Further, drawing on Kobayashi's (1994) negation of essentialism in conceptualising people's identities, she construes reflexivity as a process of self-construction rather than self-discovery:

"if the process of reflexivity changes what is being reflected upon, then there is no 'transparent' self waiting to be revealed" (Rose, 1994, p. 313).

Yet, I concur with Archer that the limited nature of reflexivity does not refute its role in the creation of individuals as discrete persons with unique sets of concerns and patterns of commitments. The failure of absolute reflexivity – that which "assumes a transparently knowable self separate from its transparently knowable context" (Rose, 1994, p. 314) – does not entail the failure of reflexivity as an essential ability of all humans to deliberate upon themselves and their place in the world. The reflexive thought *is* inherently fallible and human judgement *does* run the risk of unacknowledged conditions, misinterpreted situations, and faulty conclusions; yet, in this very fallibility lies potential for progressive evolution of one's knowledge of the private self and the social world. As Archer points out, people often realise that they have got their priorities wrong, or that their chosen commitments are, in fact, unsustainable, or that they come at too high a cost – only to engage in a new round of reflexive deliberations geared towards a better understanding of self and its contexts. This idea of socially situated, inherently fallible, but always potentially corrigible, reflexivity is at the heart of my perspective on becoming and being an ethical consumer. However, to operationalise this concept of reflexivity and unlock its explanatory potential in relation to individual actions and social phenomena, it is necessary to justify the explicit and implicit ontological assumptions upon which it rests.

A realist social ontology: towards a stratified reality

The view of reflexivity as an emergent human power and a mechanism mediating the relationship between individuals and the social world presupposes a particular social ontology, one in which structure and agents are seen as ontologically distinct but dialectically related strata, each with its unique, irreducible, and causally efficacious properties. The previous chapter has cleared away the flawed and stifling ontological presuppositions of agency-focused and socio-centric perspectives precluding an understanding of the fundamental relationship between consumers and their contexts. It is my aim, in the remainder of the chapter, to fill this cleared space with a model of social reality that accommodates both antecedently existing structures, which constrain and enable individual actions, *and* agents that have causal powers in their own right and that play an active part in the shaping of the social world. Critical realism provides a philosophical framework allowing to effectively meet these ontological demands.

A prominent critical realist, Margaret Archer, has developed an important body of work that directly addresses the long-standing controversy concerning the relationship between structure and agency. Archer's contributions to debates about social reality and change are highly relevant to the goals of this book particularly because she examines both how structure impinges upon agency and how agents respond to structural conditioning, the latter having been somewhat neglected by sociologists, including critical realists. In the third part of her trilogy, devoted to the problem of structure and agency (*Culture and Agency*, 1988; *Realist Social*

Theory, 1995; *Human Being: The Problem of Agency,* 2000), Archer redresses this imbalance by engaging in extensive theorising about agential features, properties, and powers. An attempt to define the nature of both structure and agency represents the central element of Archer's social theory, and for good reason. A robust conception of structure and agency, as Archer points out, is a prerequisite for understanding the interplay between individuals and their environments: "how structures are variously held to influence agents is dependent upon what 'structure' and 'agency' are held to be" (Archer, 2003, p. 1). Her analysis of the ontological and conceptual assumptions of some of the more influential sociological theories, such as those developed by Giddens and Bourdieu, brings to light the lamentable tendency to ascribe causal autonomy to *either* structure *or* agency, or altogether refuse to acknowledge their ontological irreducibility and analytical distinctiveness. This leads to three different, but equally deficient approaches to conceptualising social reality or, in Archer's terms, the fallacies of downward conflation, upward conflation, and central conflation. In the case of downward conflation, causal powers and capacities are granted solely to structure, while agents are denied causal autonomy in the face of the overbearing influence of structural and systemic forces and are seen as not just dependent upon, but completely subordinate to their social contexts. Where upward conflation occurs, causal power is monopolised by agents while structure, understood merely as an aggregate consequence of individual intentions and actions, is denied the potential to exert causal effects in relation to agency. In the previous chapter, I have shown how a reductionist understanding of social reality as shaped by the causal effects of either structural (upward conflation) or agential (downward conflation) properties precludes exploring the fundamental relationship between consuming agents and their contexts.

The defect of conflation may take yet another form, one where structure and agency are seen as mutually constitutive entities. This has major consequences for sociological explorations of social phenomena: to conceptualise structure and agency as co-constitutive is to render them and, by implication, their properties and powers ontologically inseparable and, most importantly, analytically indistinguishable from each other. Such an approach makes it impossible to even begin to understand either how structure influences motivations and behaviours of agents or how agents use their causal powers to mediate structural effects. Giddens' theory of structuration (1984) and Bourdieu's (1984) theory of habitus are prime examples of theoretical frameworks that attempt to transcend the antinomy between structure and agency by rejecting their ontological and analytical dualism. From a critical-realist point of view, however, this means relating structure and agency "at the cost of their analytical integrity, disabling the capacity to capture either" (Maton, 2008, p. 61). This unhelpful ambivalence manifests itself clearly in Bourdieu's attempts to define the personal through the social:

> Persons, at their most personal, are essentially the personification of exigencies actually or potentially inscribed in the structure of the field or, more precisely, in the position occupied within this field.
>
> (Bourdieu & Wacquant, 1992, p. 44)

And the individual through the common:

> Personal style ... is never more than a deviation in relation to the style of a period or class so that it relates back to the common style not only by its conformity ... but also by the difference.
>
> (Bourdieu, 1977, p. 86)

The merging of the personal and the social, which stems from Bourdieu's wanting to deprive agential subjectivity of a self-sustaining ontological footing, drives his theory right into the trap of central conflation, where one is left at a loss as to where to start in unpicking the agential and structural properties and exploring their respective contributions to social outcomes. Falling prey to this ontological fallacy is human reflexivity, for the ability of agential subjectivity to reflect upon social objectivity presupposes clear demarcation lines between agents (the deliberating subjects) and their objective circumstances (the objects of deliberation). To deny individuals the capacity for reflexive deliberations upon the self and the world is to deprive them of the ability to define their values, concerns, and commitments, to relate them to the surrounding contexts, and to decide on appropriate ways and conditions of living. On this account, the potential of agents to mediate the influence of structures upon their preferred courses of action is lost in advance, and with it their ability to make independent contributions to the making and remaking of the social world.

Such are the implications of Bourdieu's logic of practice, which replaces conscious deliberation on the part of self-aware individuals with non-reflexive workings of habitus and intentional actions by commitment-driven subjects with intuitive "feel for the game", thus, essentially depriving people of the capacity to intervene in the social world to bring about desired change. Indeed, how can anyone become aware of his or her dispositions and induce disjunctions between habitus and the field, if habitus is developed tacitly and subconsciously through socialisation and experiences, and if "principles embodied in this way are placed beyond the grasp of consciousness, and hence cannot be touched by voluntary, deliberate transformation, cannot even be made explicit" (Bourdieu, 1977, p. 75)? Further, if the function of habitus is to ensure the routine adjustment of subjective and objective structures, if, by default, "social agents ... come to gravitate towards those social fields (and positions within those fields) that best match their dispositions and to try to avoid those fields that involve a field–habitus clash" (Maton, 2008, p. 58), then how do people come to challenge the social order? As Sayer points out, habitus "makes it impossible to understand how anyone could react against and resist at least some parts of their habitat" (2005, p. 31).

The theory of habitus, which dissolves personal identity and human agency into social relations and structures to the point of denying people awareness of their own dispositions and capacity to reflexively choose their courses of action, ultimately leaves us with the "world where behaviour has its causes, but actors are not allowed their reasons" (Jenkins, 2002, p. 97). Archer insists that "we should never be satisfied with these forms of conflationary theorizing"

(1995, p. 4), for all they produce are ontologically deficient models of social reality which either see "the 'parts' dominate the 'people'" (Archer, 2000, p. 1), or allow "the 'people' to orchestrate the 'parts'" (Archer, 2000, p. 1), or collapse together what should be viewed as two distinct realms of the social world. None of these models provides solid ground for a nuanced analysis of the individual and contextual determinants of ethical consumer practices and identities because, as Bhaskar puts it, "on Model I there are actions but no conditions; on Model II conditions but no actions; on Model III no distinction between the two" (2010, p. 77). In her trilogy on structure and agency, Archer develops a way out of the conflationary state of modern social theory. Following a general critical realist approach to conceptualising reality, she makes the case for the kind of ontology that neither reduces the social world to just one dimension of the individual versus social dichotomy, nor blends structure and agency into "an amalgam whose properties and powers are completely interdependent and ineluctably intertwined" (Archer, 2007, p. 41).

Archer argues strongly for ontological and analytical dualism, suggesting that individual agents and social structures need to be understood and treated as two separate yet interdependent levels of stratified social reality, none of which has ontological supremacy over the other, or is granted a monopoly over causal power. On her account, both individual agents and social structures have their own distinctive causally efficacious properties and powers which continually interact with each other but cannot be reduced to one another. Archer characterises structural and agential properties as "emergent", for they "come into being through social combination ... they exist by virtue of interrelation" (1998, p. 192). This idea underpins the principle of emergence, which refers to the arrival of novel properties arising out of existing properties as a result of a continuous interaction between structure and agency. According to Archer, structural emergent properties (SEP), which include distribution of resources, interests, roles, doctrines, and ideologies, are a part of socio-cultural systems, which precede individuals and are ontologically irreducible to the subjective beliefs and actions of the current generation of people; hence, they are objective, pre-existing, and relatively autonomous from agents. However, personal emergent properties (PEP), such as self-awareness, reflexivity, intentionality, and personal identity, have subjective or, to use Searle's (1998, p. 42) term, "first-person" ontology. This is because individuals' thoughts, desires, intentions, and self-concepts, whilst they have full ontological status, exist only when and as experienced by people and are "ineluctably tied to the subject" (Archer, 2003, p. 37). In that sense, Archer argues, personal emergent properties are "both objectively real and subjective in nature" (Archer, 2003, p. 36). It is due to their differing ontology that agential and structural properties cannot be conflated or reduced to one another: the objectivity of social structures and the subjectivity of agents are "two causal powers that are irreducibly different in kind and make relatively autonomous contributions to social outcomes" (Archer, 2003, pp. 1–2). The distinctiveness and irreducibility of agential and structural properties are precisely what leads Archer to draw a clear line of demarcation between structure and agency:

social realism ... accentuates the importance of emergent properties at the level of both agency and structure, but considers these as proper to the strata in question and therefore distinct from each other and irreducible to one another ... Irreducibility means that the different strata are separable by definition precisely because of the properties and powers which only belong to each of them and whose emergence from one another justifies their differentiation as strata at all.

(1995, p. 14)

The structure–agency dualism and the principle of emergence lie at the core of Archer's theory of social change, which she refers to as the "morphogenetic" approach. The term "morphogenetic", as Layder (2005, p. 264) points out, reflects an attempt to capture the interplay between structure and agency:

By calling her social theory 'morphogenetic', Archer is drawing attention to the fact that the ordered forms (the 'morpho' part of the term) that society takes have their genesis (the 'genetic' part of the term) in human agency, just as social beings have their genesis in social forms.

According to Archer, the process of morphogenesis occurs in a three-stage cycle that repeats itself in perpetual circles and affects all levels of social reality, from political-economic systems to organisations to individual actors and their identities. The morphogenetic cycle begins with social structuration, which refers to the idea that all agents are born into a world of pre-existing social structures and act from positions and situations "which are not of their making yet which condition much of what they can make of them" (Archer, 1998, p. 375). These socially situated agents continuously interact with their environments as they engage in various actions and practices, which corresponds to the "social interaction" phase of the morphogenetic cycle. Through the intended and unintended consequences of their on-going social activity, agents constantly shape and reshape prior structural relations thereby inducing the elaboration (morphogenesis) or reproduction (morphostasis) of the initial contexts. Which of the two possible outcomes will ensue depends on the nature of social interaction: transgressive agential activity engenders the morphogenesis of the social world; whenever agents choose to conform to the dominant social order, morphostasis sets in. Archer's morphogenetic approach closely echoes the transformational model of social activity put forward by Bhaskar (1979), whose conception of social change also emphasises the pre-existence of social structures and their irreducibility to the activities of agents who are influenced by them:

it is no longer true to say that human agents *create* [society]. Rather we must say; they *reproduce* or *transform* it. That is to say, if society is already made, then any concrete human praxis or, if you like, act of objectivation, can only modify it; and the totality of such acts *sustain* or *change* it.

(Bhaskar, 2010, p. 60, emphasis in original)

Both approaches conceptualise human activity as consisting in the elaboration of pre-given structural properties by intentional and causally efficacious agents. Both, therefore, are varieties of a realist meta-theory that provide a fundamentally temporal and historical understanding of social processes by allowing to distinguish events, situated in space and in time, which "initiate or constitute ruptures, mutations, or generally transformations of social forms" (Bhaskar, 2010, p. 60). On a realist conception of the social and the individual, "temporality is not an option but a necessity" (Archer, 1998, p. 375), for it is entailed both by the pre-existence of social structures and their temporal (and temporary) autonomy from the doings of the current generation of agents. Realist statements about the temporal priority, relative autonomy, and causal efficacy of structural properties in relation to agents, open up opportunities for sociological investigations of the dynamics of the relationship between structure and agency and their relative contributions to social processes and outcomes.

A general critical-realist social theory with its central tenet that "the causal power of social forms is mediated through social agency" (Bhaskar, 1979, p. 26), correctly identifies structure and agency as the necessary components of any robust explanation of social change. This alone, however, tells us nothing about the nature of the mediating process or how it occurs. To fully capture and understand the dialectical relation between structure and agency, Archer argues, one will have to explore both how structural and cultural properties bear upon subjects *and* how socially conditioned agents respond to structural influences to either reproduce or transform their contexts. By foregrounding the subjective emergent property of reflexivity, the morphogenetic account of social change allows it to fulfil these analytical demands. Carrying forward a realist emphasis on the reality and causal efficacy of agential properties, Archer attributes to reflexivity the key role in mediating the relationship between structure and agency: "how people reflexively deliberate upon what to do in the light of their personal concerns has to form *a* part of a mediatory account" (Archer, 2003, p. 15, emphasis in original). This is because the conditioning – both constraining and enabling – effects of structural properties can only reveal themselves in relation to the intentions and actions of agents, for "constraints require something to constrain, and enablements something to enable" (Archer, 2003, p. 4). Thus, unless subjects entertain or embark on a particular course of action – and for that they have to exercise their reflexive capacity – the potential of structure to facilitate agential projects or impede bringing them to fruition remains unfulfilled.

Conversely, every attempt by reflexive subjects to advance their projects through the constant flux of objective enablements and constraints involves recourse to agential ability "for internal deliberation upon external reality". The contribution of human reflexivity to personal and social outcomes does not depend on whether and to what extent people's evaluation of their subjective states and objective conditions is accurate and complete; what matters is that it is through such reflexive deliberations that individuals form their subjective vision of how it would be best to pursue their projects from the specific contexts in which they are placed. The function of reflexivity is not to produce an omniscient and omnipotent actor free from determination by structural forces, but to enable individuals to consider themselves in relation to their environments and, in light of this knowledge, work out a subjectively satisfying and objectively sustainable way to act and to be. As Archer points out, "a social influence can itself be immune to what people think about it, and yet what they make of it reflexively can profoundly influence what they do about it" (2003, p. 20). In virtue of individuals' capacity to evaluate the contextual feasibility of their desired commitments, social structures can exert enabling or constraining effects upon agential courses of action through mere anticipation: the perceived ease of fulfilling a particular project is likely to strengthen the subject's resolve to embark on it, while an expectation of failures and obstacles may well prevent him from going ahead.

It is precisely because social conditioning is "a process that involves both objective impingement and subjective reception" (Archer, 2003, p. 5) that it can never be completely devoid of contributions made by agential subjectivity. This makes reflexivity an integral part of the morphogenetic cycle at all of its key stages, from structural conditioning (because structural powers depend for their activation on agents' intentions or attempts to engage in particular actions), to social interaction (since it is through their reflexively conceived projects that agents come into contact with social structures), to structural elaboration (for it is in consequence of their reflexive ability that agents can mediate the enabling and constraining influences of social contexts upon their preferred courses of action and transform the dominant social order). Agential reflexivity, therefore, is just as indispensable to the process of social change as the presence of structure itself, as Archer (2003, p. 7) makes clear in the following quote: "the effect of these structural and cultural causal powers is at the mercy of two open systems; the world and its contingencies and human agency's reflexive acuity, creativity and capacity for commitment".

A clear acknowledgement of the role of embedded reflexive individuals in the processes of social change is a key achievement of the morphogenetic approach. Its pronounced emphasis on the ontological integrity and causal efficacy of structure and agency clears away both methodological collectivism, with its tendency to explain all social phenomena solely through reference to structures and cultures, and methodological individualism favouring explanations entirely in terms of the people involved. By conceptualising social structures and agents as dialectically related whilst conceptually irreducible and analytically distinct strata of reality and reflexivity as a subjective power mediating between them, Archer accomplishes the goal of relating structure and agency, the objective and the subjective, the social and the personal in a way that avoids the pitfalls of reductionist and conflationist frameworks. Her social theory enables the production of a bilateral account of the interplay between structural forces and agential actions. Such integration of structure and agency into a unified story is crucial, for

> we cannot account for any outcome unless we understand the agent's project in relation to her social context. And we cannot understand her project

without entering into her reflexive deliberations about her personal concerns in conjunction with the objective social context that she confronts.

(Archer, 2003, p. 131)

The morphogenetic approach makes important statements about the nature of social reality, namely, that societies are dependent for their continuous existence on the people who make them up, that both structures and agency are inherently transformable, and that there exists a continuous dialectical and temporally complex relationship between structural properties and agential activity. In the following quote, Archer (1995, p. 1) amplifies these ideas:

> Firstly, [society] is inseparable from its human components because the very existence of society depends in some way upon our activities. Secondly ... society is characteristically transformable; it has not immutable form or even preferred state. It is like nothing but itself, and what precisely it is like at any time depends upon human doings and their consequences. Thirdly, however, neither are we immutable as social agents, for what we are and what we do as social beings are also affected by the society in which we live and by our very efforts to transform it.

The belief in the inherent transformability of both social structures and individual agents informs one of Archer's most important analytical projects, namely, the production of a sociological account of the process of personal change. The morphogenetic approach emphasises that the processes of change for structure and agency unfold in closely interrelated ways, and that reflexivity acts as the key driving force behind both. This is because as a personal emergent power, reflexivity has causal efficacy in relation to both social contexts and individual agents, who, in virtue of their reflexive capacity, induce structural changes while themselves being changed in the process. Archer's theory thus takes the traditional critical-realist account of social change a step further by shedding light on the mediating process between structure and agency and its transformative effects for both parties involved. The conceptual and analytical achievements of the morphogenetic approach have serious implications for sociological explorations in the field of consumer research, which should take very seriously indeed the important questions put forward by Slater (1997, pp. 172–173):

> How can we investigate the social meaning of things, needs and uses without reducing them either to omnipotent social structures (semiotic codes, grids of social classification generated by the social order itself, the structures of commercial capitalism) or regarding them as socially unconstrained, indeterminate, open, as a space of self-determined activity so free that it looks increasingly like the space of the sovereign liberal consumer?

Guided by Slater's appeal, I set out to provide a socially attuned account of ethical consumer practices and identities, one that gives insight into the world of subjective meanings underlying consumer behaviours, whilst taking note of the wider contexts in which individuals' consumption projects unfold. Having presented Archer's social theory, I now want to relate it more closely to the subject matter and aims of this book so as to more explicitly delineate the benefits of a general critical realist theory and the morphogenetic approach for the field of consumption research.

Critical realism has rich – and so far, untapped – potential for developing a more nuanced, fine-grained perspective on ethical consumption and consumer behaviour more broadly. Previously, I have shown how a skewed social ontology renders agency-focused and socio-centric paradigms conceptually and analytically unfit for capturing the interplay between consuming agents and the wider social contexts. Conversely, critical realism with its commitment to preserving the ontological integrity and analytical distinctiveness of agents and structure yields an effective theoretical and methodological framework for producing such an account. A realist social ontology accommodates both prior social structures, the causal properties of which impinge upon consumers, and shape their choices, habits, and practices, and causally efficacious agents, who respond to social forces and, depending on the nature of their responses, reproduce the dominant materialistic subjectivities and lifestyles or defy the positions and roles thrust upon them by the modern consumer society. By asserting the distinctiveness, irreducibility, and causal autonomy of structural and agential properties – the key ontological conditions that need to be satisfied for the interaction between structure and agency to become possible and open to sociological enquiry – critical realism provides a way of understanding how "structures and agents combine" (Archer, 2003, p. 8) to define consumption practices and behaviours. Importantly, it reinstates the capacity of individual consumers to challenge, change, and adapt existing social relations and contexts to their own preferred ways of provisioning, consuming, and disposing of goods.

The model of the reflexive, intentional, and causally efficacious agent put forward by Archer explains why "consumption practices are neither passively structured, nor "inevitably conformist" (Slater, 1997, p. 148), and it is precisely such agents who are – must be – the effective cause of the changes in consumption patterns and trends. Ethical consumption, which entails transgressions and modifications of the dominant attitudes and behaviours, calls for the subjects who are the ultimate authors of their practices and have conscious mastery of them, who, "even if they come from the old *héritiers*, are highly aware that they must reflexively select, suppress and supplement features from their inherited repertoire of routines" (Archer, 2007, p. 49), and whose practices "entail creative, reflexive thought about what courses of action do *constitute* mastery in and of the new context" (ibid., emphasis in original). As the next chapter will show, becoming and being an ethical consumer requires awareness of one's own values, concerns, and commitments, which are more than internalised social structures and facts. The processes underlying such self-understanding are contingent on agential ability to reflexively deliberate upon the self and its relationship to the world and repeatedly reassess this relationship in light of ongoing contextual changes. It is this capacity for reflexive self-scrutiny and self-monitoring that "enables us to be the authors of our own projects in society" (Archer, 2003, p. 34) and that accounts for our constant reinterpretation and reshaping of the social world and the personal self.

Finally, a critical-realist view of social reality as inter-subjective, inter-dependent, and inter-relational (Donati & Archer, 2015), and individuals as inherently reflexive and normative beings whose relationship to the world is one of concern (Archer, 2007), successfully accommodates human emotionality and normativity – the two stumbling blocks which the rational-choice theorist fails to negotiate. By conferring ontological status on the unobservable, yet real, entities and processes that profoundly affect agential actions, such as emotions, concerns, and reflexive deliberations, a critical-realist framework allows us to effectively account for the altruistic and selfless aspects of ethical consumer activities.

By drawing on the strengths of Archer's social theory, this book develops a sociological account of ethical consumer practices and identities that puts the figure of an individual consumer – decentred if not altogether displaced by the sceptics – into the foreground, whilst avoiding replicating the caricature portraits of consumers as freely choosing, all-knowing actors in possession of unbounded reflexive or rational powers. Reflecting well-established criticism of the extended reflexivity thesis and responding to a call for a more situated understanding of agential reflexivity, I reframe ethical consumption as a structurally conditioned, yet subjectively conceived, practice of socially embedded, yet reflexive, active, and intentional agents. My aim is to analyse and explain the emergence of the ethical consumer identity while not losing sight of the relationship between agential subjectivity and structural objectivity. In the next chapter, I apply Archer's morphogenetic theory of personal identity to ethical consumption and explore how the concept of reflexivity can help to make sense of the intricate inner process underlying the formation of consumers' ethical self.

References

Archer, M. (1988). *Culture and Agency. The Place of Culture in Social Theory*. Cambridge, UK: Cambridge University Press.

Archer, M. (1995). *Realist Social Theory: The Morphogenetic Approach*. Cambridge, UK: Cambridge University Press.

Archer, M. (1998). Realism and morphogenesis. In Archer, M., Bhaskar, R., Collier, A., Lawson, T., & Norrie, A. (Eds.), *Critical Realism: Essential Readings* (pp. 356–381). London, UK: Routledge.

Archer, M. (2000). *Being Human. The Problem of Agency*. Cambridge, UK: Cambridge University Press.

Archer, M. (2003). *Structure, Agency, and the Internal Conversation*. Cambridge, UK: Cambridge University Press.

Archer, M. (2007). *Making Our Way through the World*. Cambridge, UK: Cambridge University Press.

Barrera, M. (1986). Distinctions between social support concepts, measures, and models. *American Journal of Community Psychology*, *14*(4), 413–445.

Beck, U. (1992). *Risk Society*. London, UK: Sage Publications.

Beck, U. (1994). The reinvention of politics: towards a theory of reflexive modernization. In Beck, U., Giddens, A., & Lash, S. (Eds.), *Reflexive Modernization* (pp. 1–55). Cambridge, UK: Polity Press.

Bhaskar, R. (1979). *The Possibility of Naturalism. A Philosophical Critique of the Contemporary Human Sciences.* Brighton, UK: Harvester.

Bhaskar, R. (2010). *Reclaiming Reality.* London, UK: Taylor & Francis.

Bourdieu, P. (1977). *Outline of a Theory of Practice.* Cambridge, UK: Cambridge University Press.

Bourdieu, P. (1984). *Distinction.* Cambridge, MA: Harvard University Press.

Bourdieu, P., & Wacquant, L. (1992). *An Invitation to Reflexive Sociology.* Chicago, IL: University of Chicago Press.

Bradford, S. (2012). *Sociology, Youth and Youth Work Practice.* Basingstoke, UK: Palgrave Macmillan.

Donati, P., & Archer, M. S. (2015). *The Relational Subject.* Cambridge, UK: Cambridge University Press.

Layder, D. (2006). *Understanding Social Theory.* London, UK: Sage Publications.

Frayn, M. (1999). *Headlong.* London, UK: Faber & Faber.

Giddens, A. (1994). Living in a post-traditional society. In Beck, U., Giddens, A., & Lash, S. (Eds.), *Reflexive Modernization* (pp. 56–109). Cambridge, UK: Polity Press.

Hindess, B. (1990). Analyzing actors' choices. *International Political Science Review, 11*(1), 87–97.

Jenkins, R. (2002). *Pierre Bourdieu.* London, UK: Routledge.

Kobayashi, A. (1994). Coloring the field: gender, "race", and the politics of fieldwork. *The Professional Geographer, 46*(1), 73–80.

Martin, R. (1998). *Self-concern: An Experiential Approach to What Matters in Survival.* Cambridge, UK: Cambridge University Press.

Maton, K. (2008). Habitus. In Grenfell, M. (Ed.), *Pierre Bourdieu: Key Concepts* (pp. 49–65). London, UK: Acumen Press.

Popper, K. (1972). On the theory of the objective mind. In Popper, K. (Ed.), *Objective Knowledge: An Evolutionary Approach* (pp. 153–190). Oxford, UK: Clarendon Press.

Rose, G. (1997). Situating knowledges: positionality, reflexivities and other tactics. *Progress in Human Geography, 21*(3), 305–320.

Sayer, A. (2005). *The Moral Significance of Class.* Cambridge, UK: Cambridge University Press.

Schechtman, M. (2005). Personal identity and the past. *Philosophy, Psychiatry, & Psychology, 12*(1), 9–22.

Searle, J. (1998). *Mind, Language and Society: Philosophy in the Real World.* New York, NY: Basic Books.

Shaw, D. (2007). Consumer voters in imagined communities. *International Journal of Sociology and Social Policy, 27*(3/4), 135–150.

Slater, D. (1997). *Consumer Culture and Modernity.* Cambridge, UK: Polity Press.

Wouters, C. (1986). Formalization and informalization: changing tension balances in civilizing processes. *Theory, Culture & Society, 3*(2), 1–18.

3 Ethical consumption as a reflexive life project

It is not that "the unexamined life is not worth living", but rather that it is unliveable.

(Archer, 2000, p. 220)

My aim in this chapter is to place ethical food consumption within the broad framework of reflexivity and, more specifically, reflexive self-production and to apply this theoretical construct to elucidate the process and mechanism of the formation of ethical consumer identities. Chapter 2 has provided what I hope represents a compelling defence of reflexivity as an essential human property. Here, then, I allow myself to take the ontological status of reflexivity for granted and focus instead on exploring its role in the process of personal change. I introduce Archer's theory of personal identity and its central concepts before arguing the case for reflexivity as a key force behind the development of ethical consumer practices and identities. Through this theoretical prelude, I intend to pave the way for the interpretation of ethical consumption as a subjective life project of reflexive and purposeful agents who create and enact their identities through a continuous pursuit of their ultimate moral, environmental, and social concerns.

In pursuit of identity: reflexivity and the internal conversation

Archer (2007, p. 4) defines reflexivity as "the regular exercise of the mental ability, shared by all normal people, to consider themselves in relation to their (social) contexts and vice versa". According to her, this capacity for internal deliberation upon external reality is a personal emergent property stemming from our inevitable involvement in the world and, more specifically, our relationships with its three different orders – the natural, the practical, and the social, or discursive. Our incessant interactions with these three worldly realms give rise to three correspondingly different types of concerns: those over physical wellbeing, practical achievements, and social worth. These are ineluctable concerns, Archer (2000, p. 198) argues, for they can neither be sidestepped nor ignored due to the very nature of human life:

> All persons have to confront the natural world and … their embodiment ineluctably confers on them concerns about their physical well-being … Performative concerns are unavoidably part of our inevitable practical engagement with the world of material culture … Participation in the social realm entails concerns about self-worth which cannot be evaded in this discursive environment.

Since none of our concerns can simply be put aside, we are faced with a pressing need to achieve "a liveable balance within our trinity of inescapable concerns" (Archer, 2000, p. 221), that is, to ensure that our relationships with the practical, natural, and social orders are both subjectively satisfying and objectively sustainable. However, concerns are not only what we worry about, but also what we care about and want to achieve in light of a deep sense of commitment. From Archer's perspective, it is our ultimate concerns – what we choose to care about most in life – that defines what kind of person we are: "which precise balance we strike between our concerns, and what precisely figures amongst an individual's concerns is what gives us our strict identity as particular persons" (2000, p. 221). Thus, Archer understands a human being as essentially a "being-with-this-constellation-of-concerns" (2007, p. 87). The concepts of emotional elaboration, internal conversation, and reflexivity are central to Archer's account of how individuals arrive at their ultimate concerns. Each type of concern, she explains, gives rise to a correspondingly different cluster of emotions – emotions, as it were, are commentaries upon human concerns. The role that emotional reactions are assigned to play in relation to concerns is far from trivial; in fact, Archer considers emotions to be "central to the things we care about and to the act of caring itself" (2000, p. 194), for they indicate the presence of concerns, incite actions, and fuel commitments. The link between emotions, concerns, and actions is a necessary one since, as Archer (2007, p. 231) contends,

> It is not possible to have a genuine concern and to do nothing about it. When normal people express concern at all – as opposed to sympathy or empathy, both of which are compatible with remaining a bystander – it is usually accompanied by an attempt to do something about it.

It is human emotionality, Archer (2000, p. 225) argues, that acts as a source of the "shoving power to achieve any ends at all". However, the desired equilibrium between three different sets of concerns, which burden subjects with often conflicting demands, cannot be achieved through simply settling our lives by basic affective reactions – instead, it calls for the refinement of emotionality beyond primitive responses or biologically prescribed standards, such as choosing high-calorie foods to appease the sensation of hunger. This is what Archer refers to as emotional elaboration – an inner process via which subjects revise their original emotional reactions to the matters and situations of concern and re-evaluate them in light of their other pressing concerns with a view to defining their ultimate priorities (resisting the temptation and foregoing the cake in favour of a healthier snack because staying fit is a higher priority than indulging one's cravings, to continue the above example). Emotional elaboration forms an integral part of the inner life of every human being whose relationship to the world is one of concern, since the concerns which we designate as our ultimate and come to identify with serve as a lens through which we evaluate all our subsequent interactions with external reality: "our commitments represent a new sounding board for the emotions" (Archer, 2000, p. 242).

The subjective power that enables and propels this internal work is reflexivity, which arises precisely out of the inescapable necessity of all individuals to simultaneously interact with the three orders of the world, confront the various concerns emerging from them, and attend to their emotional imports. Reflexivity, therefore, is understood both in terms of our essential mental capacity and its continuous exercise, necessitated by the very fact of our having to constantly navigate external reality in pursuit of a satisfactory way of living. The medium through which reflexivity operates is internal conversation, conceived of by Archer as an on-going, cyclic process of inner self-talk in which people engage in order to attain and sustain an adequate balance between their competing concerns. The full cycle of internal conversation is summarised by Archer in the following sequence: discern > deliberate > dedicate, which reflects the key steps subjects go through as they: first, identify their nascent concerns (discern); second, evaluate and prioritise them in terms of their emotional appeal, moral worth, and practical feasibility (deliberate); third, reject or subjugate certain concerns while endorsing others as their ultimate concerns – those that they deem to be most important, with which they feel they can live, and which they are prepared to turn into life-long commitments (dedicate). At the end of this process, individuals not only determine what they care about most in life but, by doing so, they also define what kind of person they are: "it is these acts of ordering and rejection – integration and separation – that create a self out of the raw materials of inner life" (Frankfurt, 2007, p. 170).

The last step in the process of establishing a coherent personal identity is to develop a project, a commitment, around one's ultimate concern(s) and realise it through appropriate actions and practices. These internal commitments must be fine-tuned to the external contexts in which people are placed and from which they will be pursuing their ultimate concerns. The inevitable embeddedness of subjective life projects in objective reality is what renders the reflexive effort so indispensable: recourse to reflexivity cannot be evaded if objective enablements and constraints to agents' preferred courses of action are to be successfully negotiated and dealt with. This reflexive imperative is also a constant: the ability for internal deliberation upon external reality always stays in demand, since individuals have to continuously re-evaluate their hierarchy of concerns and reconsider their current commitments in light of constantly changing subjective states and objective conditions. I will come back to this point and discuss it extensively in Chapter 7, where the interplay between structure and agency and its role in shaping ethical consumer practices and identities will be explored in-depth.

For now, let me return to our self-creating agent, who is making her or his way towards a distinct personal identity through the following sequence of steps:

<Concerns → Projects → Practices>.

These are the key landmarks on the road paved by the subjective powers of reflexivity and leading individuals to an understanding of what matters to them most of all and how they should therefore live their lives. According to Archer (2000, p. 241), personal identity emerges out of a subject's recognition of her ultimate concerns and ensuing commitments achieved through the internal conversation: "the self and its reflexive awareness have been continuous throughout the conversation, but on its completion the self has attained a strict personal identity through its unique pattern of commitments". Following Archer's account, the underlying mechanism of personal identity formation can be summarised in the following way:

- There is a continuous and necessary relationship between human beings and the world which gives rise to a range of different concerns;
- There emerges an emotional connection between a subject and an object of concern – prompting relevant affective responses;
- These affective responses alert individuals to the presence of concerns and supply the urge-for-action;
- This triggers reflexivity – a subjective power that is ontologically real, causally efficacious, and consequential for personal and social outcomes. Its medium is internal conversation; key activities are discernment, deliberation and dedication; and its goal is to define one's ultimate concerns and a suitable way of living them out;
- As a result of this ongoing reflexive endeavour there develops a distinct personal identity defined by an individual's unique combination of concerns, reflected in her moral commitments, and actualised through suitable practices.

Archer's theory provides a socially attuned account of the inner psychological process responsible for producing one's distinct personal identity. What makes it so useful and relevant to the stated aims of this book might not be immediately obvious to the readers. The benefits of applying Archer's theory to the study of ethical consumer practices and identities will, I hope, become increasingly clear as my narrative continues to unfold. For now, let me stress two key points. First, unlike many existing accounts of identity that are largely descriptive, Archer provides an explanatory account of identity formation and unravels the mechanism underlying this very complex, subtle process. Another very important distinguishing feature of her proposed model is that it offers a detailed insight into the "raw materials" – primitive emotions and incipient concerns – out of which personal identity is produced, and elucidates the inner psychological, both mental and emotional, workings via which it is brought into being, while also not losing sight of the contextual embeddedness of identity work. It is on this critical-realist ground, where the interplay between the inner self and the outer world comes to the forefront, that I want to bring the analysis of the ethical consumer identity.

Ethical consumer – unravelling the relationship between commitments and concerns

Viewed through the lens of Archer's theory, the phenomenon of ethical consumption receives a new framework, the one in which the subjective meanings and identity work surrounding ethical consumer pursuits fully come into view. Flowing from it is a representation of ethical consumption as a subjective project, a life-organising and identity-shaping commitment at which individuals reflexively arrive through the following steps in the internal conversation with themselves:

1   Via emotional elaboration and reflexive deliberations, agents embrace the ethics of consumption as their ultimate concern thereby defining what kind of persons they are and how they should live their lives;
2   The reflexive powers are then channelled into developing a project, usually involving lifestyle changes of varying magnitude, through which subjects' moral concerns can be met; this newly intended *modus vivendi* has to be adequately accommodated within the given subjective and objective conditions of their lives;
3   Agents realise their projects of ethical consumption and, concomitantly, actualise their ethical consumer identities through what they (subjectively) perceive as morally acceptable consumption choices and behaviours. Within this framework of interpretation, every act of ethical consumption is conceived of as a tangible outcome of subjectively devised projects of reflexive and self-aware agents and a visible manifestation of their identity-defining moral concerns.

An important caveat needs to be made before taking this line of thinking about ethical consumption further. The idea of human beings using their deliberative powers to arrive at that course of action which best aligns with their subjective concerns might invite comparisons with the model of a man upheld by the proponents of the Rational Choice Theory – a consistently rational agent in pursuit of his or her own self-interest. Yet, while the "Concerns → Projects → Actions" procedure may superficially resemble the process of rational decision-making performed by a goal-oriented actor, the two represent fundamentally different and ultimately contrasting approaches to interpreting human behaviour. The key distinction lies in the fact that the internal conversation of a morally concerned agent, very much unlike the cost/benefit analysis of a preference-driven chooser, is emotion-driven and value-motivated. Reason, no doubt, plays a key role in the process: after all, individuals always have to assess their potential life projects in terms of not only their subjective desirability, but also contextual feasibility, and agents' decisions as to whether or not to pursue a particular commitment inevitably depends on their ability and preparedness to pay its associated costs. Yet, the rationality that enters the scene of the internal conversation is not instrumental rationality (Zweckrationalitat) of a utility-seeking actor, but is value-rationality

(Wertrationalitat) of a subject who treats values as ends in themselves. Archer is emphatic of the difference between the two which she explores in detail in *Making our Way Through the World* (2007). The emotional and normative dimensions of the internal conversation, she argues, cannot be reduced to value-stripped rationalisations, for they refer to the things that we care about most deeply in our lives: "right judgment stands in opposition to motivation by self-interest, idleness, self-aggrandisement, convenience and so forth" (Archer, 2007, p. 300).

A sharp contrast between preferences and concerns creates an unbridgeable gap between a preference-driven rational actor and a concern-motivated normative agent. While preferences represent a vehicle for achieving a specific goal (since whenever a rational man acts on his preferences he does so in order to advance his personal interests), concern-inspired commitments are valuable in their own right for they are not a means to some further end, but an end in themselves. Archer illustrates the point through the following analogy: "someone does not forgo a blood transfusion in order to be a Jehovah's Witness: the forgoing is an expression of being one" (2000, p. 86). This expressive aspect of the relationship between means and goals is absent from the rational-choice account of human behaviour, which assumes that individuals' actions are reflective solely of their preference schedule, and that their means is nothing more than a rationally selected instrument for achieving a desired end. While preferences are purely rational, commitments are inherently affective since the ability to pursue a project beyond one's own self-interest requires deep emotional involvement with its underlying concerns (Archer, 2000).

This pronounced emphasis on emotionality and normativity renders Archer's view of an agent highly relevant to the task of explaining ethical consumer behaviour. Her account of the relationship between human concerns, emotions, and commitments provides an effective conceptual guide to understanding the inner forces and processes behind the self-production of an ethical consumer. What requires further explanation, however, is how and why people develop concerns over food ethics in the first place. As the psychologist, Jonathan Haidt, notes, the human mind has potential to become sensitive to a variety of concerns; yet, only some out of many actual and possible matters of importance develop into fully fledged concerns, while others never become part of our personal moral matrix. It is, therefore, natural (and in the context of the stated aims of this book – imperative) to wonder how and why specific ethical concerns – environmental, social, or those pertaining to the wellbeing of animals – arise and acquire such prominence on a person's moral horizon so as to become a defining part of her or his self-identity. The question is not the one to evade if the promise of the book – to provide the most complete picture possible of the mechanism and process of an ethical consumer's coming-into-being – is to be fulfilled.

This newly posed intellectual challenge brings me to the work of Christian Coff, directly addressing the issue of food ethics. In an effort to shed light on the origins of the *Taste for Ethics* exhibited by modern consumers, Coff (2006) develops a theory that purports to explain the emergence of consumer sense of moral responsibility for the choice of food. Analysing consumer morality in the contexts of traditional and modern economies, he draws a distinction between short-range ethics – those which apply to people's immediate geographical and temporal contexts and which for a long time used to guide consumer relationship with food – and long-range or distance ethics – those which exceed the boundaries of one's "here and now" and which have become increasingly relevant to the present-day world. This idea reveals a paradoxical tension within the global food system. The ever-growing spatial and temporal gap between consumption and production processes – a distinctive feature of the modern food industry – on the one hand clearly calls for the ethics of distance, for the implications of people's diets now extend far beyond their local surroundings. On the other hand, however, it precludes their very emergence: denied insight into the world of food manufacturing, consumers remain oblivious to the ethical problems it harbours and devoid of the sense of responsibility for their day-to-day eating decisions. Yet, the main challenge, Coff argues, arises from the fact that the ethics of distance are deeply tied to the ethics of closeness: to become sensitive to the ethical impact of food production stretching far and wide across the globe, consumers must first experience it "in the local and in the present" (Kemp, 1997, p. 99, cited in Coff, 2006, p. 99). Such experiences can be obtained either personally – a visit to a slaughterhouse is one such example – or vicariously through second-hand accounts, whether oral, written, or visual. In both cases, consumers obtain what the author refers to as a "glimpsed experience" – a direct or mediated insight into production practices and their concomitant ethical problems, which prompts the diffusion of consumer sense of moral responsibility over longer distances in space and in time. Coff's account does not fall short of explaining the ethics-inducing effect of glimpsed experiences. He suggests that the following mechanism is at work here: once exposed to the gaze of yet unaware consumers, the history of food production becomes a "hi-story" – a story woven around the revealing experience which becomes part of consumers' own biographies and which comes alive every time they encounter something that reminds them of what they have "glimpsed". The food itself becomes such a reminder – "a silent document" referring to the spatially and temporally absent conditions of its production and morally disturbing aspects thereof. Such bringing of the absent into the present, Coff argues, is key to enabling people to expand their ethical vision beyond the immediacy of their own locales. This theoretical conjecture finds support in empirical research: in a study of vegans, McDonald (2000) develops an analogous concept of "catalytic experiences" to describe events and encounters that have been central to the subjects' adoption of vegan lifestyles.

While Coff sets out to explain how consumers become sensitised to concerns over food ethics, his proposed theory can serve as a useful point of departure for illuminating the ways in which moral issues are taken up in the consumption domain more broadly. Lying implicit in Coff's thinking about consumption ethics are allusions to Marx's idea of commodity fetishism, which also emphasises the inability of consumers to appreciate the true, non-monetary and non-utilitarian, value of commodities as a consequence of a profound severance between goods and the meaningful relations of their production. Hudson and Hudson (2003, p. 417) provide a clear illustration of this lamentable condition:

Under commodity capitalism, the social, environmental, and historical relations that go into the production of a commodity are hidden. When a person wanders through the grocery store or shopping mall, what they see are the characteristics of the commodities themselves—the attractiveness of the packaging, the cut of the fabric, perhaps the lifestyle associations stapled on by marketing departments, and, of course, the price. In this sense, the commodity has a life of its own, completely divorced from the process by which it was created. It becomes not a result of production on which people have worked under a wide variety of more or less acceptable conditions but an entity unto itself, with characteristics of its own.

In light of the theory of commodity fetishism, the idea of glimpsed experience acquires a deeper meaning and increased significance for understanding the relationship between the consumer and the consumed. Endowed with the potential to reveal the world of meaning behind commodities through providing a channel into the realm of production, glimpsed experiences become a means of combating commodity fetishism. By exposing the socio-historic and environmental relations invested in a product, the glimpsed experience transforms a fetishised commodity into a meaningful object of consumption. The dispelling of the aura of commodity fetishism is key to reviving consumers' sense of responsibility for the moral effects of their purchase decisions and, consequently, instigating commitments to ethical lifestyles. This conjecture finds support in the claims about the potential of ethical labels, packaging, and promotion materials to enlist individuals into projects of ethical consumption by restoring to view the conditions of production and their impact on people and the environment. Adams and Raisborough ascribe such potential to discourses surrounding fair-trade advertising which, the authors argue, undermine "the commodity fetishist lynchpin of the consumer capitalist psychic economy" (2008, p. 1172) through disclosure of exploitative relations of production. Their argument closely echoes Allen and Kovach's (2000) discussion of the role of organic labelling in reducing the objectification of natural-social relationships and weakening commodity fetishism. Similarly, Brown and Getz (2008, p. 1,188) suggest that ethical labels can offer "at least a partial antidote to the commodity fetish", while DuPuis (2000) goes as far as to claim that consumer reflexivity is turned on precisely upon reading the label.

By suggesting a plausible answer to the question about the origins of consumer concerns over food ethics, Coff adds a missing element to the theory of ethical consumer identities that I set out to develop. I find his account compelling not least because it fits in well with this book's critical-realist agenda and explanatory strategy. Coff's proposed explanation for the emergence of the sense of moral responsibility in consumers is in essence mechanism-based: the idea of glimpsed experience does not merely describe events that serve as tipping points for consumers' ethical engagement, but it also points to the causal paths in which these events bring about changes in the moral consciousness of consuming agents. Coff identifies the underlying causal links between glimpsed experiences and consumers' moral conversion and sheds light on the process through which this causal relationship comes about: direct or mediated encounters with the world of production give rise to the narratives which unfold the ethical meaning of products and, through becoming part of consumers' own life stories, engender their sense of moral responsibility in relation to food choice. It is because of this deep engagement with causality that Coff's account works so well in tandem with the critical-realist approach underpinning this book and can be used effectively to fill the outstanding gap in my proposed theory. I integrate complementary insights derived from the works of Archer and Coff to suggest the following mechanism for the formation of the ethical consumer identity:

- Through glimpsed experiences of moral issues surrounding production and consumption, ethical concerns become woven into a subject's moral matrix;
- A continuous emotional commentary sensitises the subject to the presence of the emergent concerns and provokes a genuine urge to address them through actions;
- This triggers a reflexive conversation during which the subject reviews her prospective ethical commitments in light of her other concerns and evaluates them in terms of their emotional appeal, moral worth, and practical feasibility;
- By designating concerns over the ethics of consumption as her ultimate concerns, the subject attains the ethical consumer identity;
- Finally, consumers actualise and sustain their ethical self-image by engaging in ethically perceived consumption practices and activities.

Laid out above is the fundamental mechanism revealing a chain of causally linked events, which lead to the transformation of an individual from an ordinary to a morally concerned consumer and give rise to a range of subjective consumption commitments that create and sustain the ethical self. Later, I will provide evidence from empirical research with ethical food consumers to demonstrate the presence of this mechanism in ethical consumption, its relevance and explanatory potential in relation to the questions posed in this book. I will begin by presenting an in-depth analysis of events and encounters that played a key role in inciting respondents' ethical concerns – this section will especially emphasise the value of the concept of glimpsed experience for understanding how the consumer journey towards the ethical self begins to unfold. Despite the study's fairly small sample, I have a wide range of experiences to report on: visiting pastures, farms and abattoirs, watching a documentary or a fantasy movie, biting into a meat pasty – all these seemingly trivial episodes were described by participants as highly consequential and relevant for their sense of food ethics. These are just some examples of the various situations that can set into motion the causal chain leading to changes in consumers' moral mind-sets. The list is indicative, not exhaustive: the mechanism may be triggered by many other kinds of events in so far as they offer consumers a glimpse of production-related issues that are worthy of moral concern.

I do not hereby argue that the particular incidents that had life-changing consequences for the individuals who took part in my research always lead to identical outcomes. I believe we can safely assume that many people have lived through the same experiences without the same effects: for example, many people have been to a factory farm or an abattoir and continue to eat meat (as is the case with one of my interviewees), albeit presumably the degree of confidence in the moral validity of such behaviour will vary from person to person. Such counterfactuals, although not hard to find, do not undermine Coff's account of consumer ethics. The fact that not all people who have been or, indeed, are continuously exposed to the harsh realities of the modern food industry (e.g. those overseeing or work-ing in sweatshops and factory farms) develop commitments to ethical consump-tion, does not make the causal relationship between glimpsed experiences and the emergence of moral concerns any less real. While the idea of glimpsed experience points to a general mechanism that might be used for explanatory purposes in a wide array of scenarios, when it comes to understanding variations and excep-tions in outcomes, knowing more about the particular conditions of any given case becomes key. Archer's theory goes a long way toward identifying the condi-tions which make this general mechanism applicable in specific cases. These are the same conditions that need to be satisfied for any concern to be designated as ultimate, namely,

- that it is of sufficient emotional import to appear among the subject's incipi-ent concerns (the "discern" phase in the cycle of internal conversation);
- that it passes the test of reflexive scrutiny and earns a place on the subject's list of concerns which are deemed both subjectively satisfying and objec-tively sustainable (the "deliberate" phase); and
- that it gets promoted to the very top of the subject's hierarchy of concerns thus becoming her ultimate concern – that which defines the person's identity and guides her subsequent life-course trajectory (the "dedicate" phase).

The process of consumer moral conversion may hit an impediment and come to a halt at any of these stages. Glimpsed experience may fail to engender an emo-tional reaction of sufficient magnitude to begin with – after all, it must be granted that some people are just less compassionate, empathetic, or sensitive than others. An incipient concern might, upon deeper reflection, be rejected as incongruent with or less appealing than the subject's other concerns (for reasons that need not necessarily be selfish, as we will see later). Further along it may also occur that the subject, however deeply concerned, will have to withdraw from the current commitment due to no longer being able to meet its concomitant costs.

As a life project, ethical food consumption engenders complex relations between the agent and the three worldly realms: in the natural realm, there is a direct link between food, body, and health; in the practical order, reducing one's negative impact as a consumer requires certain knowledge, resources, and skills; and in the social sphere, commitment to ethical eating needs to be reconciled with the responsibilities and obligations that arise from various other positions and roles assumed by the subjects in the course of their lives. Thus, under-standing the nature and incidence of the process of consumer moral conversion requires consideration of its contextual differentiation. The effects of the natural, practical, and social contexts in which agents are placed is the factor that can make a decisive difference to consumers' ability and preparedness to engage with consumption ethics. How will new habits affect their health? Do they have access to knowledge and practical resources needed to initiate and sustain the desired behaviour change? What status effects will the new lifestyle bring and how will it impact their friendships, romantic, and workplace relationships? These are not merely rhetorical questions – they demand concrete answers, and the answers will determine whether and to what extent ethical consumer intentions will translate into actions. How and with what consequences for the self-consumers respond to these worries and negotiate the uncertainties and dilemmas that arise on their way towards ethical consumption is the question I take up in the second part of the book.

References

Adams, M., & Raisborough, J. (2008). What can sociology say about FairTrade?: class, reflexivity and ethical consumption. *Sociology*, *42*(6), 1165–1182.

Allen, P., & Kovach, M. (2000). The capitalist composition of organic: the potential of markets in fulfilling the promise of organic agriculture. *Agriculture and Human Values*, *17*(3), 221–232.

Archer, M. (2000). *Being Human: The Problem of Agency*. Cambridge, UK: Cambridge University Press.

Archer, M. (2007). *Making Our Way through the World*. Cambridge, UK: Cambridge University Press.

Brown, S., & Getz, C. (2008). Privatizing farm worker justice: regulating labor through voluntary certification and labeling. *Geoforum*, *39*(3), 1184–1196.

Coff, C. (2006). *The Taste for Ethics: An Ethic of Food Consumption*. Dordrecht, the Netherlands: Springer.

DuPuis, E. (2000). Not in my body: BGH and the rise of organic milk. *Agriculture and Human Values*, *17*(3), 285–295.

Frankfurt H. (2007). Identification and wholeheartedness. In Frankfurt H. (Ed.), *The Importance of What We Care About: Philosophical Essays* (pp. 159–176). Cambridge, UK: Cambridge University Press.

Haidt, J. (2012). *The Righteous Mind*. New York, NY: Pantheon Books.

Hudson, I., & Hudson, M. (2003). Removing the veil?: commodity fetishism, Fair Trade, and the environment. *Organization & Environment*, *16*(4), 413–430.

McDonald, B. (2000). "Once you know something, you can't not know it". An empirical look at becoming vegan. *Society & Animals*, *8*(1), 1–23.

Part II
Studying the ethical consumer

4 Studying consumption

A realist approach

Understanding consumer culture is a matter of social rather than textual analysis, not an enterprise of reading but rather of explaining and accounting.

(Slater, 1997, p. 148)

In this chapter, I outline and discuss the methodology, design, and process of my enquiry into ethical consumer practices and identities. I begin by demonstrating the relationship between my ontological commitments, epistemological position, and my approach to planning, implementing, and evaluating research. I provide the rationale for my chosen research tools, in-depth interviews and direct observations, and consider their limitations and strengths – not merely according to textbooks, but also, and most importantly, in terms of how they played out in practice and what challenges and opportunities they gave rise to. Finally, I reflect on the ways in which my subjectivity and positionality as a researcher and consumer influenced the research relationships, process, and outcomes, and describe how through the use of various reflexive and discursive techniques, I sought to understand and control, to the extent possible, my personal impact on the production of knowledge about ethical consumers.

Epistemological approach: towards the interpretivist paradigm

There is an inextricable link between our ontological position, epistemological orientation, and the ways which we consider most suitable for conducting and assessing research. In my case, the ideas that social reality is made up of two separate strata, agents and structure, and that each of these strata possesses its own distinct properties and causal powers, which together determine social phenomena, processes, and outcomes, played a key role in dictating my epistemological standpoint. So did the beliefs, to which I pledge allegiance throughout the book, that all human beings have the capacity for internal deliberation upon external reality, which they exercise during constant internal self-dialogues, that all people, *ipso facto*, have rich and complex inner lives, and that subjective, unobservable, and intangible entities such as concerns, emotions, and reflexive deliberations constitute the raw materials, "the stuff" of the inner life of each person.

Underpinned by these ontological beliefs, my intention to reveal the underlying drivers and determinants of ethical consumption behaviour through exploring the interplay between agential subjectivity (individuals' concerns, commitments, and ensuing identities) and structural objectivity (intangible but real causal effects that structural properties exert on agential intentions and actions) rules out empiricism as an appropriate research paradigm. Concerned with testable predictions about concrete, readily observable, and easily measurable phenomena (Goulding, 1999), empiricism falls short of providing effective tools for exploring subjective meanings, motivations, and dispositions attached to the behaviour of individuals. Since no empirical test can capture or measure the inner workings of people' minds or their subjective relationship to objective reality, for both are impossible to penetrate and explore merely by direct perception of facts, with no recourse to ideas or concepts, empiricist approaches have little to contribute to our understanding of the generative mechanisms of human concerns, commitments, and practices.

An enquiry well attuned to the subtleties and complexities of moral behaviour should acknowledge that "an appeal must be made to something non-observable" (Brown, 1982, p. 234), and that "the imperative to explain is sometimes an imperative to posit theoretical entities" (Brown, 1982, p. 234), which are more than "hand-maidens to the larger goal of prediction" (MacDonald, 2003, p. 554). It calls for a method which is sensitive to and takes account of the underlying subjective meanings and motivations of agential practices and which recognises that people's internal states – their feelings and thoughts – cannot be deduced or learned simply from observing their external behaviour. These requirements provide a compelling reason for approaching the study of ethical consumption from an interpretivist perspective, which emphasises subjective understanding over objective knowledge and offers potential to shed light on the ways in which people think, feel, and behave in given contexts (Marsh & Furlong, 2010). Interpretivist epistemology aligns well with critical realism, for both accept that many valid accounts of a single phenomenon may exist and that "there is no possibility of attaining a single, 'correct' understanding of the world, what Putnam (1999) describes as a 'God's eye view' that is independent of any particular viewpoint" (Maxwell, 2012, p. 5). Critical realists, while insisting that objective reality exists independently of our minds, understand knowledge as a social development involving many points of view and types of meaning: "all knowledge is thus 'theory-laden', but this does not contradict the existence of a real world to which this knowledge refers" (Maxwell, 2012, p. vii). Such a position, argues Maxwell, has been widely accepted as "a commonsense basis for social research" (2012, p. 6). Frazer and Lacey (1993, p. 182) also defend the compatibility of a realist ontology and an interpretivist epistemology: "even if one is a realist at the ontological level, one could be an epistemological interpretivist … our knowledge of the real world is inevitably interpretive and provisional rather than straightforwardly representational". But perhaps the most insightful and lucid explanation of the relationship between knowledge and reality is given by the anthropologist Barth (1987, p. 87) in his study of indigenous communities in Papua New Guinea:

Like most of us, I assume that there is a real world out there—but that our representations of that world are constructions. People create and apply these constructions in a struggle to grasp the world, relate to it, and manipulate it through concepts, knowledge, and acts. In the process, reality impinges; and the events that occur consequently are not predicated on the cultural system of representations employed by the people, although they may largely be interpretable within it.

While I contend that human reflexivity plays a key role in shaping social processes and outcomes, I recognise that the study of social reality cannot be limited to exploring people's experiences and understandings of it, for there is an important distinction between the ontological realm of what exists and the empirical domain of what can be experienced and observed. In the words of Spencer, "there remain ontological questions about society since much of society lies outside the realm of thought itself" (2000, n/p). This is a major, crucial point to recognise if we are to go beyond simply reporting or re-describing people's experiences and perceptions of social reality to capture more complex, deep-lying causal processes and interactions which contribute to the shaping of this reality over time.

Interpretive paradigms, such as phenomenology and hermeneutic phenomenology, have affirmed their relevance to consumer research and are widely used in the field of consumer studies (e.g. Ahuvia, 2005; Arnold & Fischer, 1994; Cherrier, 2006; Thompson, Locander, & Pollio, 1989; Thompson, Pollio, & Locander, 1994). The value of interpretive analysis for exploring consumer practices and identities lies in its ability to go beyond a narrow focus on consumer-buying behaviour and bring to light the experiential and meaningful aspects of consumption acts and activities. Holbrook and Hirschman (1993) are emphatic about the relevance of interpretivism for the study of consumer behaviour, while McQuarrie and McIntyre (1990) and Thompson, Pollio, and Locander (1994) specifically argue for the adoption of a phenomenological position in consumer research. In a study of married women's consumption experiences, Thompson, Pollio, and Locander (1994) evidently demonstrate the potential of phenomenological analysis to provide valuable insights into consumer behaviour. While both phenomenology and hermeneutic phenomenology offer a means for exploring social phenomena through subjective experiences of individuals and groups (Kafle, 2013), phenomenological research tends to be largely descriptive and concentrates on the structure of experience, while hermeneutic analysis places a premium on interpretation and is ultimately concerned with the underlying meanings of people's experiences and their effects on individuals and social forms (Laverty, 2003). More specifically, the hermeneutic model of understanding allows the contextualisation of "the meaning of particular life events ... within a broader narrative of self-identity" (Thompson et al., 1994, p. 451). In an article discussing the application of hermeneutical framework in consumer research, Thompson, Pollio, and Locander (1994, p. 448) highlight how a hermeneutic interpretation explicates the personalized meanings by which consumers understand the characteristics of their (perceived) actual identities, ideal identities, and undesired identities (Markus and Wurf, 1986) and the ways in which these identity perceptions (and their underlying meanings) are manifested in everyday consumption activities.

Importantly, hermeneutics takes a distinctly different stance towards "preunderstanding", that is, a set of prejudices and prejudgements that every researcher inevitably brings to bear upon his or her study (Arnold & Fischer, 1994). In contrast to other research paradigms, including phenomenology, hermeneutics sees pre-understanding as what facilitates rather than hinders interpretation; consequently, researchers are encouraged to capitalise on their subjective ideas and understandings rather than trying to set them aside before embarking on research. In consumer studies, researchers' pre-understanding stems both from their position as academics who possess theoretical knowledge about consumer behaviour and from their personal experiences as consumers. Thus, the application of hermeneutics to the study of consumption allows researchers "to draw more consciously, critically, and powerfully on their own [pre-]understanding of the everyday phenomena that we study" (Arnold & Fischer, 1994, p. 66). By doing so, they can re-experience, recognise, and rethink what participants felt or thought (Bleicher, 1980), thereby achieving understanding at intellectual, emotional, and moral levels (Betti, 1990).

These specific aspects of a hermeneutic approach, namely, its pronounced emphasis on the subjective motivations and meanings attached to individuals' behaviours; its acknowledgement of the potential of researchers' personal experiences to serve as a valuable source of data in understanding the phenomenon under study; and its capacity to yield insight into the intellectual, emotional, and normative dimensions of human conduct make it apt for exploring the relationship between ethical consumer practices and identities. Importantly, hermeneutic phenomenology accommodates what Morse (1994) refers to as the constants of qualitative research – comprehension, synthesising, and theorising. As a research paradigm, it is ultimately geared towards the development of a social theory through revealing the common structures of people's experiences and is, therefore, well-suited for identifying the underlying causal mechanisms of social processes and phenomena.

Detailed research design

Research ethics. This research project was reviewed and approved by the Research Ethics Committee of the University of Leeds. All respondents received electronic and paper forms outlining the details and conditions of the research to ensure that they understood the nature of the study and their right to withdraw from the project at any time. The participants were assured that their identities would be protected through the use of self-chosen pseudonyms and anonymisation of sensitive details, and that all personal data relating to them would be destroyed in a secure manner when no longer required. Prior to the start of the fieldwork, I requested the respondents' signed consent for the following: to take written notes during observations, to audio record the interviews, to transcribe the interview data verbatim, to include anonymised data and interview excerpts in any publications that may result from the research.

Selecting participants – a targeted approach. To accentuate the qualitative nature of my research, I will talk about the process of participant recruitment using the term "selection" rather than "sampling", which has become too tightly associated with quantitative survey designs (Emmel, 2013; Maxwell, 2012; Stake, 1995).

My key criterion in selecting participants was their sense of self as ethical subjects, that ethical is "a fundamental part of who they are, how they think and feel, and ultimately affects how they consume" (Connolly & Prothero, 2008, p. 213). The project was advertised via posters and flyers distributed in ethically oriented places and sites such as natural food shops, vegetarian and vegan eating establishments, and ecological housing developments. This was considered a legitimate way to access ethical consumers because, drawing on Archer's arguments that "subjects acquire their personal identities through the constellation of concerns that they endorse" (2012, p. 22) and that "it is not possible to have a genuine concern and to do nothing about it" (2007, p. 231), I assumed that self-identities of ethical consumers must be tied to concerns about the moral, environmental, and social implications of their lifestyle choices and that these concerns must translate into concrete actions, that is, decisions about where to buy groceries, where to eat, and where to live. Far from random, these assumptions are a manifestation of internal powers – the particular ideas, concepts, and theories which researchers are guided by and which profoundly influence the recruitment process (Emmel, 2013).

My final sample consisted of nine self-perceived ethical food consumers (five females and four males between 29 and 64 years old) from the North of England. Among different forms of ethical consumerism, I chose to focus on ethical eating due to the close relationship between food and identities exemplified in a vast body of literature (Diner, 2001; Fischler, 1988; Lang & Heasman, 2004; Warde, 1997). Moreover, food consumption encompasses a wide range of moral issues, including social justice, animal rights, and environmental protection, and can therefore offer an insight into a variety of moral concerns that drive consumer engagement in ethical practices. Table 4.1 summarises the main characteristics of the study subjects. As it shows, most of the participants, although not all, were either vegans or vegetarians; their similar eating behaviours, however, were driven by different motives, including the condemnation of the killing of animals, the promotion of animal welfare, a respect for the environment, a commitment to social justice and human rights. Among the respondents, there were five committed organic consumers and five active supporters of fair trade. Thus, despite its limited size, the sample was sufficiently diverse to allow insight into a wide range

Table 4.1 Main characteristics of the study participants

Pseudonym	Sex and age	Occupation	Ethical practice	Ethical concerns
David	Male, 33	PhD student	Vegetarian Consumer of organic, local, and fair-trade products	Environment and climate change Social justice and human rights Animal welfare
Darren	Male, 36	Activist	Vegan	Animal life and welfare
Lucy	Female, 48	Occupational therapist	Vegetarian	Animal life and welfare
Solveig	Female, 29	PhD student	Vegan	Animal life and welfare
Mary	Female, 64	Retired	Consumer of organic, local, and fair-trade products	Environment and climate change Social justice and human rights Animal welfare
Maggi	Female, 62	Retired	Vegan Consumer of organic, local, and fair-trade products	Animal life and welfare Environment and climate change Social justice and human rights
Manasi	Female, 31	Engineer	Vegetarian Consumer of organic and local products	Environment and climate change Social justice and human rights
Jason	Male, 31	PhD Student	Consumer of organic, local, and fair-trade products	Environment (biodiversity) Social justice and human rights
Joe	Male, 29	Call centre agent	Vegan Consumer of local products	Animal life and welfare Environment and climate change Social justice and human rights
Lila	Female, 34	Editor	Vegan Consumer of organic, local, and fair-trade products	Animal life and welfare Environment and climate change Social justice and human rights

of moral concerns attached to ethical consumption practices and their meanings in the context of the respondents' identities.

As far as the sample size is concerned, using a small number of respondents is a general guideline in qualitative and, particularly, hermeneutic phenomenological research (Creswell, 2012). Polkinghorne (1989) specifies a range of 5 to 25 respondents, Dukes (1984) recommends focusing on 3 to 10 subjects, while Boyd (2001) considers 2 to 10 participants to be sufficient for a hermeneutic enquiry. Such small numbers are justified by the purpose of qualitative research, which is not to yield generalisable findings but to focus on "information-rich" cases, that is those "from which one can learn a great deal about matters of importance" (Patton, 2001, p. 242). From this viewpoint, the empirical data generated through interviews, observations, and informal discussions with nine ethical consumers proved exceptionally nuanced, multi-layered, and rich; in fact, the desired level of depth and detail in the analysis and interpretation of the respondents' accounts might have been hard to achieve with a larger number of interviewees.

Data production: the rationale, limitations, and benefits of my chosen research methods

Consistent with my deeply anti-empiricist approach to sociological enquiry, I abandon the term "data collection" with its implicit assumption that social "facts are lying about waiting for the researcher to spot them" (Ramazanoglu & Holland, 2002, p. 154). Following Ramazanoglu and Holland, I understand the process of generating empirical data in qualitative research in terms of "data production": far from endorsing the social constructionist view of reality, this term subtly emphasises that "information gathered by the researcher is produced in a social process of giving meaning to the social world" (Ramazanoglu & Holland, 2002, p. 154).

Interviews. My primary research tool was semi-structured interviews integrating a life-history approach with elements of in-depth phenomenological interviewing. Such interviewing strategy has been previously used in consumer studies and proved to serve well the researcher's objectives (see, e.g. Fournier's (1998) study of consumers' relationships with brands). I chose this combination of methods to generate two types of information: (1) participants' accounts of their experiences of becoming and being ethical consumers situated within the larger narratives of their lives; and (2) respondents' views and understandings of the relationship between identity, consumption, and ethics.

My overall goal was to understand how, that is through which inner workings and under which external influences, individuals achieve and maintain their desired ethical consumer identities. All interviews were conducted within a time frame of two to four weeks after completion of observations, which were my complementary data production technique. I used this time to review observation and field notes and build a tentative mental portrait of each participant to ensure an effective interview process. I began each interview by asking respondents to recount their life story focusing on a particular aspect of it, namely their ethical consumption practices and the evolution thereof over time. Interviewees were then invited to reflect on the relationship between consumption and ethics and the meaning of their ethical commitments in the context of their lives as unique individuals and as consumers. When conducting the interviews, I aimed to adhere to the principle of emergent dialogue, according to which the leading role in steering the course of the conversation should be assigned to respondents (Thompson, Locander, & Pollio, 1990). While closely following the lines of the narratives told by the interviewees, I tried, as much as possible, to keep the discussion in the realm of the relevant themes, and elicit stories describing the key steps of participants' journeys towards ethical consumption. The interviews varied in duration from one to five hours, but generally lasted for around two and a half hours. I transcribed all interview data verbatim myself to facilitate data analysis by increasing the level of closeness between the interpreter and the text (Halcomb & Davidson, 2006). Although the use of verbatim quotations in academic texts is sometimes advised against (Corden & Sainsbury, 2006), I decided not to edit excerpts from interviews to further accentuate the participants' own voice and their subjective meaning-making.

Why interview? The rationale for using interviews. Hermeneutic phenomenology is non-prescriptive in relation to methods of data production and allows the use of various research techniques (van Manen, 1997). However, if one accepts, as I do, that individuals' reflexive deliberations, albeit objectively real, are ontologically subjective (I have discussed the first-person ontology of personal emergent properties in more detail in Chapter 2), one must also acknowledge the ineluctability of what Archer (2003) refers to as the "first-person perspective" – the idea that only people themselves can have a direct, unmediated access to the contents of their rich inner lives. The "first-person perspective" entails belief in the "first-person authority", which implies that individuals' self-understanding always has epistemic privilege over third-person examination. Archer (2003, p. 22) is emphatic about the impossibility of "exteriorising our interiority" and the falsity of the belief that "everything inner can be read from its public behavioural manifestations". Thus, researchers can only access – indirectly and in no other way than by means of interpretation – the inner repositories of one's authentic self-knowledge by soliciting and analysing first-person accounts.

My choice of in-depth one-to-one interviews as the main method of data production ensued precisely from these beliefs. Assuming that the data relevant for my study is contained within the perspectives of those who have direct experiences of becoming and being ethical consumers, I believe that it is only by making central the voices of participants in my research that I can hope to bring to light the underlying motivations and meanings of their commitments and practices. Being sensitive to the epistemic privilege of those being studied, I relied on first-person accounts as a key means of inquiry precisely because "these stories treat the human being and his/her mind as invaluable to understanding and explaining social behaviour" (Orbuch, 1997, p. 468). My choice of interviews as the prime research tool is, therefore, linked directly to the realist belief that reflexivity plays a key part in shaping agential concerns, commitments, and practices. In contrast, theoretical perspectives that deny agents the capacity to contribute to their own life courses in a reflexive, self-aware, and intentional manner, render people's accounts of their experiences inconsequential, for how can a person give an explanation for the actions that "are the product of a modus operandi of which he is not the producer and has no conscious mastery?" (Archer, 2000, p. 42). Such frameworks, of which Bourdieu's theory of habitus is one prominent example, make the interview method unfit for penetrating the meaning making of social actors, for it invites individuals "to reflect verbally upon matters that are inaccessible, because unconscious, and ineffable, because embodied rather than discursive" (Archer, 2000, p. 43).

As a critical realist adopting an interpretivist epistemology, I strongly concur that "we have to take the agent very seriously indeed because he or she is a crucial source of self-knowledge" (Archer, 2003, p. 33). Importantly, to examine events, relations, and processes through the subjective prism of human perception is not to deny or diminish their ontological status as objectively real social phenomena. Accepting the epistemological privilege of first-person narratives does not mean "substituting how agents take things to be for how they really are" (Archer, 2003, p. 15). By relying on people's accounts of their social life, one does not automatically reduce social reality to subjective opinions and interpretations: "the ontological status of something real is not impugned by allowing that it can be valued differently by different subjects", points out Archer (2003, p. 140). At the same time, it needs to be recognised that narratives of personal experiences cannot be seen as straightforward factual accounts – such naïve and simplistic treatment of interviews has been criticised by a number of commentators (see, e.g. Atkinson & Silverman, 1997; Gubrium & Holstein, 1995). To avoid the uncritical use of interviews in social research, we need to consider how personal narratives are produced and shared in an interview context and be aware of the inevitable limitations of the interview method, its process, and outcomes.

The limitations of the interview method. The use of interviews as a key strategy for eliciting data, which, as I argued, no one other than agents themselves can directly access and claim knowledge about, presents significant challenges that need to be explicitly acknowledged and carefully considered. The life-history approach is inherently problematic for, as Archer (2003, pp. 31–32) notes, when we ask people to recall and account for something from their past, "we are asking for attentive retrospection. This is not like taking a second look at a filed photograph; it is much more like police procedure where witnesses are asked to recall 'any detail, however trivial'". This raises several concerns. First, the question arises as to whether and to what extent we may expect people to retain memories of old experiences, which might well have gone unnoticed, and reproduce them at the investigator's request with the precision needed to inform an accurate analysis? Second, even if it is possible for a person to revive past events and imagine experiencing them again, can he also reanimate the same mental and emotional states he went through at the time? Has Lila, one of the ethical consumers who took part in my research and whom we will meet later, succeeded in "trying to go back to my younger self and see how it felt then", or has she failed in her self-reflexive endeavour, proving that "the life of the mind is fundamentally Heraclitan, for it never descends twice into the same stream" (Archer, 2003, p. 60)? Third, if James (1950, p. 234) is right, that "experience is remoulding us every moment, and our mental reaction on every given thing is really a resultant of our experience of the whole world up to date", does it then follow that whatever a person makes of his past decisions and actions can never be held as a true reflection of his former self, since our interpretation of any given moment from the past is always and inevitably refracted by all our subsequent experiences? Lawler holds precisely this view, arguing that "significance is conferred on earlier events by what comes later" (2008, p. 16), and that in telling their life stories people always engage in "'memero-politics' – a process by which the past is interpreted in light of the knowledge and understanding of the subject's 'present'" (2008, p. 18). As she contends, "it is not simply that memories are unreliable (although it is): the point is that memories are themselves social products. What we remember depends on the social context" (Lawler, 2008, p. 17). Steedman (1986, p. 5) makes the same point when she describes how memory makes the current self through the interpretation of the past:

> We all return to memories and dreams … again and again; the story we tell of our life is reshaped around them. But the point does not lie there, back in the past, back in the lost time at which they happened; the only point lies in interpretation. The past is re-used through the agency of social information, and that interpretation of it can only be made with what people know of a social world and their place within it.

It is noteworthy that my research participants, my storytellers, were themselves conscious of the inherent fallibility of human memory and the traps involved in relying on people's accounts of the past. This was best reflected in a comment made by Lila regarding her mother's recollections of Lila's childhood: "I am not sure if it is true because sometimes people kind of re-write memories to fit their ideas", she said, unwittingly fuelling concerns about the reliability of her own narrative.

Finally, if the overarching goal of my research can only be achieved by gaining knowledge that is being kept in the sole possession of my respondents, how can I ascertain the truthfulness of their accounts and ensure that no important details have been deliberately or inadvertently omitted, withheld, or misrepresented? Is it possible to rely on a person's narration of the past to understand his way towards his present self if, as Goodson and Sikes (2001, p. 6) contend, life stories are always "lives interpreted and made textual" and inevitably present only "a partial, selective commentary on lived experience" and if, as Goffman (1959) famously argued, people are actors engaged in an almost incessant process of impression management and presentation of self? These methodological hurdles, while demanding most careful consideration, need not stand in the way of a rigorous sociological enquiry or prevent attempts at gaining a genuine understanding of human behaviour. Upon deeper reflection, neither the inherent limitations of retrospective interviewing nor the focus on data that is difficult to access and impossible to validate should stand in the way of producing the knowledge I want. As argued by Lawler (2008), ultimately it does not matter whether the stories told by participants are fully comprehensive and entirely accurate accounts of their experiences (Biesta et al. (2001) are probably right in suggesting that they never are), for the value of personal narratives lies in the way authors use them in the construction of desired identities. Underlying this argument is the view of a person's self "as 'made up' through making a *story* out of a *life*" (Lawler, 2008, p. 11, emphasis in original), that is "through a series of creative acts in which she interprets and reinterprets her memories and experiences, articulated within narrative" (Lawler, 2008, p. 12). Such perspective presupposes a particular way of understanding the relationship between individuals' self-narratives and their identities, namely seeing identity as being produced during the storytelling through selecting particular memories, episodes, and experiences and assembling them within narrative. Thus, it is "not that autobiography (the telling of a life) reflects a pre-given identity: rather, identities are produced through the autobiographical work in which all of us engage every day" (Lawler, 2008, p. 13).

This theoretical position informs my methodological approach to the narratives told by the study subjects: I treat them as meaningful and intentional configuration of the various happenings and events in which participants have engaged into a coherent whole – their life story. Drawing insights from Polkinghorne (1988), I see these episodes as playing contributing roles in bringing about particular life outcomes; in the context of my research, my goal is to understand how these events interacted to produce ethical consumer identities out of the raw materials of participants' inner lives. In this analysis, the meaning of each single episode is derived from the recognition of the part it played in a systemic whole – the subject's entire life – the concerns it gave rise to, the conflicts and struggles it generated, and the direction in which it steered his or her course of action. To reiterate, these are not randomly chosen episodes, they "have a place in the plot and so they produce the narrative" (Lawler, 2008, p. 12); together, they constitute a life story which is "always the same story in the end, that is the individual's account of how she got to be the way she is" (Steedman, 1986, p. 132). Inevitably, narrated life stories rely on the current interpretation of past experiences and events, and hence "the "now" is … always present in one's story of the past" (Biesta et al., 2011, p. 9). Considering the present contexts from which individuals recount, reconstruct, and reinterpret their biographies and staying attuned to their role in the production of people's self-narratives is therefore key to producing a valid and reliable research account.

Observations. The observational method was used to assist in generating rich, detailed, and varied data and provide a background for participants' ethical consumption practices. Initially, I planned to conduct observations in the context of grocery shopping, the intention being to accompany participants on their weekly shopping trips over the course of a month. However, the original plan was subsequently adjusted to accommodate a wide variety of consumption practices, shopping habits, and forms of food provisioning in which participants engaged and which came to light during fieldwork. Thus, apart from shopping trips to different types of food retail outlets, such as mainstream supermarkets, specialty food stores, and farmer's markets, observations were also conducted during visits to food growing sites such as allotments and gardens, co-housing communities, charity shops, and online shopping. The number, frequency, and destinations of observation trips were agreed with each participant on an individual basis. Such responsiveness to the demands placed on the study design by the actual research context and the varied needs of respondents is a crucial component of a critical-realist approach to planning and carrying out qualitative research (Maxwell, 2012). Informal conversations occurring naturally during observations proved an important source of information, which contributed to my understanding of participants' ethical consumption commitments and practices. I chose not to audiotape these discussions to allow for a less formal interaction, but notes were taken of any cues arising from participants' behaviour, verbal and physical. I approached these spontaneous conversations as an opportunity for detecting potentially important themes to be further explored via interviewing. Accordingly, careful analysis of all relevant data generated through observations constituted an integral part of my preparation for formal interviews.

Why observe? The rationale for using observations. In adopting a multi-method approach, I was guided by methodological literature suggesting that drawing on different sources of evidence is a powerful strategy that can enhance the credibility of research and increase confidence in the validity of the findings (Bryman, 2008; Denscombe, 2014). Through using interviews and observations as mutually reinforcing qualitative techniques (Patton, 2001), I sought to illuminate participants' consumption practices from different angles and secure an opportunity to corroborate their accounts, thereby reducing the uncertainty of my interpretations and improving the reliability of my conclusions. However, as my understanding of the nature of the knowledge I was seeking evolved, as my focus on the real yet unobservable processes and entities has become increasingly sharper, and as the key strengths and limitations of my chosen research tools started to manifest themselves during fieldwork, I was prompted to re-evaluate the role and relevance of the observational method in my study. Most importantly, the data collected through observations was never used to verify participants' accounts. Such treatment of observational data would run contrary to my intention to explore the "non-observable" of ethical consumption – the invisible, intangible, immeasurable entities such as concerns, emotions, commitments, and the reflexive interactions between them. As a research tool, observations provide no means for penetrating the subjective meanings that people attach to their consumption behaviours, and it is only by analysing respondents' narrated accounts of their practices and experiences that I was able to achieve the desired depth of insight. I discovered, for instance, that the same consumption choices and decisions may be inspired by different, and even contrasting, concerns and commitments; that seemingly value-laden consumer activities may be completely void of ethical motives; and that the most deeply held moral principles may never manifest themselves in behaviour due to various personal and structural factors that determine consumers' ability to act upon their beliefs.

Whilst of no use as a means of comparison and contrast, observations proved absolutely critical in allowing me to approach the standards of rigour of interpretive and, more specifically, hermeneutic phenomenological research. One of the key methodological prescriptions of hermeneutic phenomenology is that individuals' experiences must be related to and understood in light of the specific "life worlds" in which they arise (Thompson et al., 1990). The aim of the researcher, therefore, is to approximate the research process to the studied experience as it is lived. Through observing participants engage in various forms of ethical consumption behaviour, from shopping for ethically labelled products in supermarkets and specialty stores to digging vegetable patches in community gardens and allotment sites, I was able to better understand the contexts in which ethical choices were made and ethical consumption activities took place and thereby achieve closeness to the real experiences of ethical consumers *as they were being lived*. Observations also created meaningful opportunities for participants to talk about their ethical consumption experiences whilst living through them, thus allowing for better "involvement of the researcher in the world of the research participants and their stories" (van Manen, 1997, cited in Kafle, 2013, p. 196). Such "bathing in the experience as it occurs" (Grbich, 2007, p. 88) played an important role in enhancing my frame of reference – the background knowledge and understanding of the phenomenon under study, a key determinant of the quality and credibility of hermeneutic phenomenological research (Thompson et al., 1994).

Finally, by spending time with participants prior to formal interviews, I was able to lay a solid foundation for building trust and rapport. Lawrence-Lightfoot and Davis (1997) highlight the importance of the amount of time spent and the frequency of encounters with respondents for achieving productive and trustful research relationships. Through extensive and varied communication, I managed to build the level of trust that is needed for individuals to feel comfortable to share their life stories with an outside person. The level of openness that characterised my relationship with participants is reflected in the following comments from Maggi: "I will tell you anything"; and Mary: "this is quite personal, but I don't mind telling you". The environment of safety and trust that was created to a large extent due to observations allowed me to fully harness the potential of my primary research method, interviewing, and achieve the desired depth and breadth of enquiry. In hindsight, what was intended as a complementary data production technique and a tool for corroboration of findings became an essential means for ensuring the quality and trustworthiness of my research.

Reflexivity in research

Being a reflexive researcher. The idea of reflexivity is absolutely central to my entire research enterprise. It is an overarching concept in my theoretical framework and the guiding principle of my methodological approach: I explored how individuals were being reflexive in consumption while simultaneously engaging in a reflexive effort myself. Both critical realism and hermeneutic phenomenology emphasise the role of the researcher as a co-producer of information and draw attention to the effects of researcher subjectivity on the analysis and interpretation of data (Goulding, 1999; Maxwell, 2012). Realist research design demands that the researcher's identity and perspectives be taken into account (Maxwell, 2012); likewise, hermeneutic phenomenological tradition requires the researcher to accept the impossibility of "bracketing", that is, suspending "all previous ontological judgment about the situation in an attempt to gain access to the common-sense knowledge and practical reasoning used by the group under study" (Goulding, 1999, p. 863). In consumer studies more specifically, Wallendorf and Brucks (1993) and Gould (1995) reject the notions of objectivity and distance and call for acknowledgement of the part played by researcher subjectivity in the process of knowledge creation.

Driven by concerns over the need to account for the subjective impact built into the process and outcomes of my work, I set out to explore how various aspects of my subjective self – my biography, personal, and professional characteristics, my unique worldview – influenced my sociological enquiry. I aimed to acknowledge and account for these "interpretive influences" (Laverty, 2003, p. 24) thereby enhancing the credibility of my study. Yet, whilst sharing the view that "rather than engaging in futile attempts to eliminate the effects of the researcher, we should set about understanding them" (Hammersley & Atkinson, 2007, p. 16), I found it challenging to work out "how, specifically, one becomes aware of this subjectivity and its consequences, and how one uses this subjectivity productively in the research" (Maxwell, 2012, p. 98). Following methodological advice, I engaged in reflexive self-analysis through introspective writing, which involved identifying and describing my assumptions, feelings, and beliefs about the phenomenon and people I studied. This self-revelatory exercise has yielded extensive notes, akin to what Maxwell (2012) refers to as "researcher identity memos" and Preissle (2008) calls "subjectivity statements". Some of these notes I used solely for personal reflection, while others were published on my personal research blog. The latter allowed me to go beyond mere private reflections about my relationship to the research and with the researched, and present them publicly to support the study conclusions and outcomes, as prescribed by Denscombe (2014). By being "open to public scrutiny and amenable to evaluation" (Denscombe, 2014, p. 284), every weekly blogpost I published contributed to my accountability as a researcher, author, and analyst. This reflexive record was intended to enhance the transparency of my study "by establishing a vantage point for critically assessing the researchers themselves, their integrity, their decisions on questions of research design, strategy, methods and theoretical framework and the data that result" (Johnstone, 2007, p. 113).

In the field, I sustained the reflexive focus on the self by making it an integral component of my relationships with participants, an essential constituent of the process of listening and understanding their narratives (Maxwell, 2012).

Like Tolman and Brydon-Miller, I strove to "bring myself knowingly into the process of listening, learning from my own thoughts and feelings in response to what [a participant] is saying in her story" (2001, p. 132) and in doing so, improve "my ability to stay clear about what my own ideas and feelings are and how they do or do not line up with [participants'] words, thus avoiding 'bias' or imposing my story over [theirs]" (2001, p. 132). I continued to maintain reflexive consciousness at the stage of data analysis to be able to monitor how aspects of my personal and professional self permeated into my interpretations and (re)presentations of participants' narratives. My reflexive effort was directed towards methodological goals: I engaged in reflexivity in the hope of yielding the best possible research outcomes – a more transparent account, more valid interpretations, more reliable knowledge.

Beyond my researcher-self: situating myself in relation to ethical consumers. Building a harmonious relationship with participants and achieving a sense of trust to allow for the free flow of information (Spradley, 1979) involved more than managing my position as a researcher and interviewer. At some point during the study, I became aware of the need to consider, and to a certain extent control, respondents' perceptions of me as a person of certain principles and beliefs. I was first prompted to think about the ways in which my personal ethics and consumer position may influence my research when I received a request from a prospective participant to refrain from sending her Microsoft Word documents as she was boycotting the corporation on ethical grounds. It was at this point that I realised that my respondents could bring their ethical identity and moral agency into our relationship, and that my handling of these sensitive issues could affect the research process and outcomes. These considerations became increasingly prominent during fieldwork which involved casual meetings and informal conversations with participants. Somewhat unexpectedly, I found that they approached me in exactly the same way as I approached them – as an emotional, reflexive, and normative human being whose relationship to the world is – *must be* – one of concern. Given the focus of my study, some of my respondents assumed that I identified as an ethical consumer and pursued an ethical diet or lifestyle myself; however, at that time I did not explicitly commit to or engage in consumption practices that fall under this category and could not, therefore, claim an ethical consumer identity. I initially assumed that this difference could create psychological distance and inhibit open dialogue with respondents, thereby posing a validity threat to my research. I was later convinced by Maxwell that we "need to avoid assuming that solidarity is necessarily a matter of similarity, and to be prepared to recognise the actual processes through which difference can contribute to relationship" (2012, p. 102). Indeed, honesty about my personal consumer behaviour contributed to the development of symmetric and reciprocal research relationships – those which "reflect a more responsible ethical stance and are likely to yield deeper data and better social science" (Lawrence-Lightfoot & Davis, 1997, pp. 137–138).

Acknowledging my lack of participation in ethical consumption was not merely a matter of being honest with those whom I expected to be honest with me, it was also a means of ensuring a comfortable space between me and my participants and avoiding the "danger" of too much rapport (Seidman, 1998). By highlighting that my interest in ethical consumption was of an academic as opposed to a personal nature, I was able to subtly accentuate my role as a researcher and achieve a balanced relationship with participants. At the same time, I made a conscious effort to ensure that my personal lifestyle and consumption choices did not disrupt the environment of comfort, safety, and trust that I had created for them. I deliberately refrained from wearing leather shoes, having regular milk with my coffee, or exposing branded items, such as the iPhone or MacBook, when meeting with my respondents. This required not only emotional sensitivity, but also careful considerations of a more practical sort, such as how to dress and even what file format to use. I found myself closely engaging with the "whole webs of signification … built up around apparently tiny clues" (Gray, 1995, p. 162), which could be detected and subjected to scrutiny by my ethically minded participants.

Overall, establishing and sustaining productive and mutually satisfying relationships with respondents proved an emotionally demanding and highly self-reflexive activity. Underlying my persistent efforts to monitor my behaviour as a researcher, and consumer, was the understanding that "the relationships that the researcher created with the participants in the research are real phenomena; they shape the context within which the research is conducted, and have a profound influence on the research and its results" (Maxwell, 2012, p. 100). In line with the standards of realist research, managing research relationships was essential to ensuring the quality of my study.

Data analysis

I approached the transcription of interviews as the first stage of data analysis. While transcribing participants' narratives, I made notes on the relevant concepts, recurrent patterns, and dominant themes. In this way, I compiled a summary sheet for each interview transcript to be added to the respective "ethical consumer case" – a detailed profile of each participant containing his or her background information, observation and field notes, verbatim transcription of the formal interview, and any other documents and materials that provided insight into the respondent's ethical consumer concerns, practices, and experiences (Joe's profile, for instance, included excerpts from his personal diary containing a reflexive account of his shopping and consumption practices). From these documents, textual data for analysis and interpretation ensued.

While the hermeneutic phenomenological tradition emphasises that the process of textual interpretation is irreducible to a set of methodological procedures (Gadamer, Weinsheimer, & Marshall, 2004; van Manen, 1997) and thus none is offered, various scholars (Arnold & Fischer, 1994; Laverty, 2003; Thompson et al., 1994) suggest that the data be analysed through a hermeneutic circle – a series of iterations between the parts and the whole of the text. In a study of the sociocultural meanings underlying consumer experiences, Thompson, Pollio, and

Locander (1994) provide detailed methodological guidelines for the application of hermeneutic phenomenology to the research process, emphasising that a thorough analysis should aim at the thematic interpretation of data through a hermeneutic endeavour – an iterative reading of the text and on-going revisions of prior interpretations in light of the constantly developing understanding of the relationship between the text as a whole and its parts. The authors distinguish between intra-textual movements, whose aim is to achieve an understanding of the text as a whole, and inter-textual iterations, which seek to establish distinctions and similarities across different texts. The purpose of this interpretation strategy is to elicit, through each subsequent reading of the text, a broader range of essential meanings until an integrated understanding of the text is achieved, at which point the hermeneutic cycle may stop (Kvale, 1996).

Broadly guided by these recommendations, I analysed the data by engaging in iterative readings of participants' personal narratives, focusing on the flow of events and looking for their antecedents, consequences, and interdependencies. I decided to refrain from the coding of the data to avoid what Maxwell (2012, p. 115) describes as "context-stripping", that is, neglecting the contextual relationships within which different data segments originally belong and which are usually lost as the result of the categorising analysis. Concerned with the actual contexts in which consumers' ethical commitments and practices emerged and unfolded, I sought to preserve the diverse and complex contextual relations around them and hence chose not to segment participants' narratives into discrete data units. Like Atkinson (1992, p. 460), I was interested in "reading episodes and passages at greater length, with a correspondingly different attitude towards the act of reading and hence of analysis. Rather than constructing my account like a patchwork quilt, I [felt] more like working with the whole cloth". For the same reason, that is, to avoid cutting up "the whole cloth" of participants' stories, I decided against using computer-assisted data analysis software and instead relied on printed copies of interview transcripts and a highlighter pen. This old paper-and-pencil method proved not only an effective, but also a more satisfying way of immersing myself in the data and connecting with participants' stories on both mental and physical levels. A small number of participants allowed me to avoid categorising the data without losing the ability to compare and find connections between interviews and reveal patterns and commonalities in respondents' narratives. I moved back and forth across different participant stories to highlight similarities and divergences in respondents' accounts of their ethical-consumer experiences first at the level of concrete, but moving further beyond the explicit – facts, events, openly stated feelings, and thoughts – to discern subtler, more abstract processes and relations captured in the data. The analysis was geared towards comprehension, synthesising, and theorising and involved ongoing revisions of previously achieved understandings and tentatively drawn conclusions. The hermeneutic circle continued until the data was rendered meaningful and turned into credible evidence which could be brought into a dialogue with my theoretical framework.

Assessing the validity of research: a realist approach

As a realist, I reject procedure-based approaches to research validity which imply that validity is assured by following methodologically rigorous procedures and rules. From a realist perspective, validity is understood not as a property of methods, but as inherent in the relationship between a research account and the things it purports to account for: "validity thus pertains to the accounts or conclusions reached by using a particular method in a particular context for a particular purpose, not to the method itself" (Maxwell, 2012, p. 130). Nor do judgements of validity apply to data, for, as Hammersley and Atkinson (2007, p. 223) point out, "data in themselves cannot be valid or invalid; what is at issue are the inferences drawn from them". According to Maxwell, assessing the validity of research is a matter of evaluating the credibility and trustworthiness of the understandings and conclusions reached by the researcher: "understanding is a more fundamental concept for qualitative researchers than validity" (2012, p. 133). Importantly, to apply the notion of validity to researcher's interpretations and understandings is not to suggest that there is only one "correct" or "objective" way of looking at the situation or problem in hand, but to highlight the importance of evaluating the relationship between the research account and the things that it intends to account for.

As a methodological framework, hermeneutic phenomenology is consistent with a realist approach to the quality of research in that it sees understanding as key to credible interpretation and theorising. The breadth and depth of researcher's "orientation" (van Manen, 1997) or "frame of reference" (Thompson et al., 1994), that is her background knowledge and understanding about the phenomenon under study, is considered an important determinant of the quality of interpretation and trustworthiness of research findings.

In the context of hermeneutic phenomenological studies, of primary concern is their interpretative validity, that is, the degree to which the researcher's conclusions are informed by understanding arising from the perspective of those being studied. Similarly, from a realist perspective "accounts of meaning must be based initially on the conceptual framework of the people whose meaning is in question" (Maxwell, 2012, p. 138). Addressing threats to interpretive validity is a challenging task, since the inner workings of people's minds can be neither observed nor directly accessed, and researchers' understanding thereof is always a matter of making inferences from respondents' own accounts. Moreover – and this is crucial given my study's ambition to uncover the underlying mechanism of ethical consumer identities – judgements of interpretative validity pertain not only to the values and beliefs which people make a conscious and deliberate use of, but also to the unconscious impulses, desires, and motives that shape their actions. It is the task of the investigator to analyse both those ideas and feelings, which subjects acknowledge as their own, and those which they may not be able or willing to recognise as important influences on their attitudes, intentions, and actions.

Whilst being aware that all understanding is inherently fallible and inevitably incomplete, I sought to maximise the validity of my research account by adhering, to the extent possible, to the standards of rigour of qualitative and, more specifically, hermeneutic phenomenological research. The latter advises researchers to enhance their frame of reference by analysing their personal experiences of the phenomenon under study, soliciting experiential descriptions of it from participants, and engaging with relevant literature to gain deeper insight and refine conclusions (Goulding, 1999). In keeping with these recommendations, I began to develop my understanding of ethical consumption through reviewing my personal experiences of it. This effort brought sparse results due to my very limited prior exposure to ethical consumption as a social phenomenon and ethical consumers as a personality type – a gap which I attribute to the specifics of the socio-cultural environment, informational context, and consumption opportunities prevailing in Azerbaijan, the country in which I was born and raised. I addressed this potential weakness in a number of ways. I subscribed to a range of ethical consumer magazines and newsletters, began to attend public events addressing relevant issues and themes, and set up a blog reflecting how my understanding of ethical consumption was evolving during the research. Further, I started to buy ethical foods (e.g. organic, fair trade, free-range) to gain direct experience of searching for, identifying, and choosing products with ethical attributes. In the process, I compiled a list of ethical labelling schemes that appeared to dominate the UK food market; this gave me an idea about the types of ethical products that people were most likely to consume, the range of ethical issues they addressed, and the kinds of moral concerns they spoke to. Finally, I used observations as an opportunity to immerse myself in the lifeworlds of ethical consumers and gain first-hand knowledge of their practices and behaviours. This allowed me to mitigate the validity threats inherent in my primary research method, interviews. As Maxwell (2012) notes, by using insights, yielded from a brief and limited interaction during the interviewing process, to make inferences about the rest of the subject's actions, feelings, and thoughts we inevitably give rise to concerns about internal generalisability of our findings, that is the degree to which we can project our conclusions about the processes studied onto those that remained outside the interview situation. Through observations, I was able to reveal aspects of participants' practices and identities that did not come to light during interviews, and in this way reduce the uncertainty of my interpretations.

Underlying my approach to the study design and implementation is the idea that the first criterion for quality in social research is "fidelity to the phenomenon under study, not to any particular set of methodological principles" (Hammersley & Atkinson, 2007, p. 7). This is especially true of qualitative research design which, in the words of Maxwell, is more of "a 'do-it-yourself' rather than 'off-the-shelf' process" (2012, p. 76). Keeping in mind that "rules of method serve us, but only to a certain point, after which they may enslave us" (Sandelowski, 1954, p. 56), I sought to preserve the "art in science" by adjusting methodological prescriptions to the demands of the actual research context and exploring the potential of established techniques to serve new purposes and offer enhanced benefits. Thus, I observed the principle of respondent-led interviewing only so far as it was possible to do so without undermining the focus on the production of knowledge relevant to the goals of the study. I deviated from my original plan to use methodological triangulation as a means of data verification: instead of using observations to validate participants' accounts of their ethical consumption practices, I treated them as an opportunity to establish rapport and lay the basis for a productive research relationship. These relationships have themselves become an integral component of the overall study design, as prescribed by realist methodology (Maxwell, 2012). In addition, observations played a key role in increasing confidence in the interpretive validity of my research account through contributing to my frame of reference and enhancing the internal generalisability of my conclusions. Similarly, I approached the process of data analysis as "an interpretive act" rather than "a precise science" (Saldaña, 2012, p. 4). I analysed the data without recourse to the commonly prescribed categorising and coding procedures, which I considered too mechanical, reductive and, most importantly, destructive for the valuable connections between data and their context, and hence for my ability to achieve a systemic view of the complex causal processes and relations captured in it. Overall, my approach to the specific techniques and procedures deployed in the study has been that of a realist researcher who assesses her methods "in terms of the actual context and purpose of their use" (Maxwell, 2012, p. 148) and the ways in which they contributed to a valid research account.

The outcome of the research process described above is a small-scale study of ethical consumer practices and identities. Although my selection of participants reflects a range of variously positioned subjects and different forms of ethical consumption, it cannot claim to be representative of all individuals who may self-identify as ethical consumers and does not produce statistical generalisations. The aim of my enquiry was "not to generalize but rather to provide a rich, contextualized understanding of some aspect of human experience through the intensive study of particular cases" (Polit & Beck, 2010, p. 1451). The value of my study lies in its potential to drive social theory, to promote an understanding of ethical consumption which goes beyond the concrete level of particular judgments and actions of individual agents and extends to the level of the causal processes and interactions underlying this complex phenomenon. It produces what Yin (1993) calls "analytic" and Seale (1999) refers to as "theoretical" generalisation – that which allows making projections about the particular mechanisms at work from one case to others. As Muys (2009, p. 43) puts it:

> The full and thorough knowledge of the particular is also a form of generalization, not in the sense of scientific induction but as a naturalistic generalization that is arrived at by recognizing the similarities of objects and issues in and out of context.

The value of my research, therefore, lies not in its capacity to "enumerate frequencies" (Yin, 2009, p. 15), but in its potential "to expand and generalize theories" (ibid.) and, more specifically, to promote new theoretical perspectives on identity formation in ethical consumption. Admittedly, my study draws on local and specific cases and thus invokes a particular set of powers giving rise to ethical consumer concerns, commitments, and identities. As Brown (2014) points out, realist researchers are limited in their ability to grasp the totality of complex systems consisting of a myriad discrete structures and powers, which determine the causes of emergence and shape the conditions of existence of social processes and forms. In synthesising the results of my analysis of local and specific ethical consumer cases into a unified theory about the generative mechanism of ethical consumer identities, I am inevitably constrained by the very specificity and locality of these cases. The particular contexts from which my participants constructed their life and identity narratives, and those from which I accessed them, inescapably limit the reach of my enquiry and draw boundaries around what I was able to reveal, explore, and explain. Yet, such context-specific research endeavours are worthwhile and necessary for, as Brown points out, "we are never going to find out about the system as a whole without local and specific enquiry" (2014, p. 118). Through exploring local and specific cases, realist researchers build towards their overarching goal: to explain social phenomena by identifying their underlying structures and mechanisms.

References

Ahuvia, A. (2005). Beyond the extended self: loved objects and consumers' identity narratives. *Journal of Consumer Research, 32*(1), 171–184.

Archer, M. (2000). *Being Human: The Problem of Agency*. Cambridge, UK: Cambridge University Press.

Archer, M. (2003). *Structure, Agency, and the Internal Conversation*. Cambridge, UK: Cambridge University Press.

Archer, M. (2007). *Making Our Way Through the World*. Cambridge, UK: Cambridge University Press.

Archer, M. (2012). *The Reflexive Imperative in Late Modernity*. Cambridge, UK: Cambridge University Press.

Arnold, S., & Fischer, E. (1994). Hermeneutics and consumer research. *Journal of Consumer Research, 21*(1), 55–70.

Atkinson, P. (1992). The Ethnography of a medical setting: reading, writing, and rhetoric. *Qualitative Health Research, 2*(4), 451–474.

Atkinson, P., & Silverman, D. (1997). Kundera's immortality: the interview society and the invention of the self. *Qualitative Inquiry, 3*(3), 304–325.

Barth, F. (1987). *Cosmologies in the Making*. Cambridge, UK: Cambridge University Press.

Betti, E. (1990). Hermeneutics as the general methodology of the Geisteswissenschaften. In Ormiston, G. L., & Schrift, A. D. (Eds.), *The Hermeneutic Tradition: From Ast to Ricoeur* (pp. 159–197). New York, NY: SUNY Press.

Biesta, G., Field, J., Hodkinson, P., Macleod, F., & Goodson, I. (2011). *Improving Learning Through the Lifecourse: Learning Lives*. London, UK: Taylor & Francis.

Bleicher, J. (1980). *Contemporary Hermeneutics: Hermeneutics as Method, Philosophy and Critique*. London, UK: Routledge & Kegan Paul.

Boyd, C. (2001). Phenomenology the method. In P. Munhall (Ed.), *Nursing Research: A Qualitative Perspective* (pp. 93–122). New York, NY: National League for Nursing Press.

Brown, A. (2013). Critical realism in social research: approach with caution. *Work, Employment & Society, 28*(1), 112–123.

Brown, J. (1982). The miracle of science. *The Philosophical Quarterly, 32*(128), 232–244.

Bryman, A. (2008). *Social Research Methods*. Oxford, UK: Oxford University Press.

Cherrier, H. (2006). Consumer identity and moral obligations in non-plastic bag consumption: a dialectical perspective. *International Journal of Consumer Studies, 30*(5), 515–523.

Connolly, J., & Prothero, A. (2008). Green consumption: life-politics, risk and contradictions. *Journal of Consumer Culture, 8*(1), 117–145.

Corden, A., & Sainsbury, R. (2006). Exploring "quality": research participants' perspectives on verbatim quotations. *International Journal of Social Research Methodology, 9*(2), 97–110.

Creswell, J. (2012). *Qualitative Inquiry and Research Design*. Thousand Oaks, CA: Sage Publications.

Denscombe, M. (2014). *The Good Research Guide: For Small-Scale Social Research Projects*. Maidenhead, UK: McGraw-Hill Education.

Diner, H. (2001). *Hungering for America*. Cambridge, MA: Harvard University Press.

Dukes, S. (1984). Phenomenological methodology in the human sciences. *Journal of Religion & Health, 23*(3), 197–203.

Emmel, N. (2013). *Sampling and Choosing Cases in Qualitative Research*. London, UK: Sage Publications.

Fischler, C. (1988). Food, self and identity. *Social Science Information, 27*(2), 275–292.

Fournier, S. (1998). Consumers and their brands: developing relationship theory in consumer research. *Journal of Consumer Research, 24*(4), 343–353.

Frazer, E., & Lacey, N. (1993). *The Politics of Community*. Toronto, Canada: University of Toronto Press.

Freeman, M. (1998). Mythical time, historical time, and the narrative fabric of the self. *Narrative Inquiry, 8*(1), 51–76.

Gadamer, H., Weinsheimer, J., & Marshall, D. (2004). *EPZ Truth and Method*. New York, NY: Bloomsbury Publishing.

Goffman, E. (1959). *The Presentation of Self in Everyday Life*. Garden City, NY: Doubleday.

Goodson, I., & Sikes, P. (2001). *Life History Research in Educational Settings*. Buckingham, UK: Open University Press.

Gould, S. (1995). Researcher introspection as a method in consumer research: applications, issues, and implications. *Journal of Consumer Research, 21*(4), 719–722.

Goulding, C. (1999). Consumer research, interpretive paradigms and methodological ambiguities. *European Journal of Marketing, 33*(9/10), 859–873.

Grbich, C. (2007). *Qualitative Data Analysis*. London, UK: Sage Publications.

Halcomb, E., & Davidson, P. (2006). Is verbatim transcription of interview data always necessary? *Applied Nursing Research, 19*(1), 38–42.

Hammersley, M., & Atkinson, P. (2007). *Ethnography: Principles in Practice*. London, UK: Routledge.

Holbrook, M. B., & Hirschman, E. C. (1993). *The Semiotics of Consumption: Interpreting Symbolic Consumer Behavior in Popular Culture and Works of Art* (Vol. 110). Berlin, Germany: Mouton de Gruyter.

Holstein, J. A., & Gubrium, J. F. (1995). *The Active Interview* (Vol. 37). Thousand Oaks, CA: Sage.

James, W. (1950). *The Principles of Psychology*. New York, NY: Dover Publications.

Johnstone, B. (2007). Ethnographic methods in entrepreneurship research. In Neergaard, H. & Ulhoi, J. (Eds.), *Handbook of Qualitative Research Methods in Entrepreneurship* (pp. 97–121). Cheltenham, UK: Edward Elgar.

Kafle, N. (2013). Hermeneutic phenomenological research method simplified. *Bodhi: An Interdisciplinary Journal*, 5(1), 181–200.

Kvale, S. (1996). *Interviews*. Thousand Oaks, CA: Sage Publications.

Lang, T., & Heasman, M. (2004). *Food Wars*. London, UK: Earthscan.

Laverty, S. (2003). Hermeneutic phenomenology and phenomenology: a comparison of historical and methodological considerations. *International Journal of Qualitative Methods*, 2(3), 21–35.

Lawler, S. (2008). *Identity*. Cambridge, UK: Polity Press.

Lawrence-Lightfoot, S., & Davis, J. (1997). *The Art and Science of Portraiture*. San Francisco, CA: Jossey-Bass.

MacDonald, P. (2003). Useful fiction or miracle maker: the competing epistemological foundations of rational choice theory. *The American Political Science Review*, 97(4), 551–565.

Marsh, D., & Furlong, P. (2010). A skin not a sweater: ontology and epistemology in political science. In Marsh, D. & Stoker, G. (Eds.), *Theory and Methods in Political Science* (pp. 1–41). Basingstoke, UK: Palgrave Macmillan.

Maxwell, J. (2012). *A Realist Approach for Qualitative Research*. Thousand Oaks, CA: Sage Publications.

McQuarrie, E., & McIntyre, S. (1990). What the group interview can contribute to research on consumer phenomenology. In Hirschman, D. (Ed.), *Research in Consumer Behavior* (pp. 165–194). Greenwich, CT: JAI Press.

Morse, J. (1994). *Critical Issues in Qualitative Research Methods*. Thousand Oaks, CA: Sage Publications.

Muys, M. (2009). *Substance Use among Migrants*. Brussels, Belgium: ASP.

Orbuch, T. (1997). People's accounts count: the sociology of accounts. *Annual Review of Sociology*, 23(1), 455–478.

Patton, M. (2001). *Qualitative Research and Evaluation Methods*. Thousand Oaks, CA: Sage Publications.

Polit, D., & Beck, C. (2010). Generalization in quantitative and qualitative research: myths and strategies. *International Journal of Nursing Studies*, 47(11), 1451–1458.

Polkinghorne, D. (1989). Phenomenological research methods. In Valle, R. & Halling, S. (Eds.), *Existential-phenomenological Perspectives in Psychology* (pp. 41–60). New York, NY: Plenum Press.

Preissle, J. (2008). Subjectivity statement. In Given, L. (Ed.), *The Sage Encyclopedia of Qualitative Research Methods* (pp. 845–846). Thousand Oaks, CA: Sage Publications.

Ramazanoglu, C., & Holland, J. (2002). *Feminist Methodology*. London, UK: Sage Publications.

Saldaña, J. (2012). *The Coding Manual for Qualitative Researchers*. London, UK: Sage Publications.

Sandelowski, M. (1994). The proof is in the pottery: toward a poetic for qualitative enquiry. In Morse, J. (Ed.), *Critical Issues in Qualitative Research Methods* (pp. 48–63). Thousand Oaks, CA: Sage.

Seale, C. (1999). *The Quality of Qualitative Research*. London, UK: Sage Publications.

Seidman, I. (1998). *Interviewing as Qualitative Research*. New York, NY: Teachers College Press.

Slater, D. (1997). *Consumer Culture and Modernity*. Cambridge, UK: Polity Press.

Spencer, N. (2000). On the significance of distinguishing ontology and epistemology. Retrieved July 18, 2017, from Hegel Summer School website, http://ethicalpolitics.org/seminars/neville.htm

Spradley, J. (1979). *The Ethnographic Interview*. New York, NY: Holt, Rinehart and Winston.

Stake, R. (1995). *The Art of Case Study Research*. Thousand Oaks, CA: Sage Publications.

Steedman, C. (1986). *Landscape for a Good Woman*. London, UK: Virago.

Thompson, C., Locander, W., & Pollio, H. (1989). Putting consumer experience back into consumer research: the philosophy and method of existential phenomenology. *Journal of Consumer Research, 16*(2), 133–146.

Thompson, C., Locander, W., & Pollio, H. (1990). The lived meaning of free choice: an existential-phenomenological description of everyday consumer experiences of contemporary married women. *Journal of Consumer Research, 17*(3), 346–361.

Thompson, C., Pollio, H., & Locander, W. (1994). The spoken and the unspoken: a hermeneutic approach to understanding the cultural viewpoints that underlie consumers' expressed meanings. *Journal of Consumer Research, 21*(3), 432–452.

Tolman, D. L., & Brydon-Miller, M. (2001). *From Subjects to Subjectivities: A Handbook of Interpretive and Participatory Action Research Methods*. New York, NY: New York University Press.

Van Manen, M. (1997). *Researching Lived Experience: Human Science for an Action Sensitive Pedagogy*. London, Ontario: Althouse Press.

Wallendorf, M., & Brucks, M. (1993). Introspection in consumer research: implementation and implications. *Journal of Consumer Research, 20*(3), 339–359.

Warde, A. (1997). *Consumption, Food, and Taste*. London, UK: Sage Publications.

Yin, R. (2009). *Case Study Research: Design and Methods*. Thousand Oaks, CA: Sage Publications.

5   Meeting the ethical consumers

I present the results of my study on ethical consumers through telling their stories. Inspired by Andrew Sayer, I deliberately move away from the approach that treats readers "more as fellow spectators of social life than as possible co-participants" (Sayer, 2011, p. 11). This attitude, as Sayer points out, is shared by many social researchers, who tend to explain social phenomena through third-person accounts of other people's behaviour and feelings without explicitly asking or even tacitly encouraging readers to evaluate presented portrayals in light of their personal life experiences.

Since this book explores issues that are close to everyone's heart – for no one can possibly live a life without ever considering the questions of morality, ethics, and how we should live – I find the urge to "address the readers as fellow participants in life" (Sayer, 2011, p. 11) especially justified. I therefore want to invite my readers to step out of the position of detached observers and, as I walk them through the life stories of nine ethical consumers, reflect on their own personal experiences as thinking and emotional beings, as moral agents who have rights and responsibilities, and as unique individuals in a relentless pursuit of their own subjective vision of the good life. To be able to assess my account of the inner and outer worlds of ethical food consumers, one does not have to be or imagine herself a proponent or connoisseur of alternative modes of consumption. To identify with the individuals who took part in my research and comprehend their moral projects and actions, one only needs to think of herself as a person whose relationship to the world is one of concern. This, I believe, should not be a difficult task for any of us, for we all have our own worries and cares which, whether we want it or not, feed into our deliberations about what to do with our life, steering us towards certain courses of action and away from the others. An exercise in reflexive self-scrutiny is all it takes for a person to discern the presence – or absence – of specific concerns in every intentional action she takes or decides to avoid.

The moment one starts to construe her own life trajectory in terms of a wider picture of her subjective concerns, understanding the behaviour of others becomes easier too. The fact that what matters to me or to you may be radically different from the concerns pursued by the people whose ethical pathways we will soon start exploring has little importance, if any at all. Regardless of the nature and roots of our subjectively developed concerns, the effects they exert on us – inherently moral, evaluative, emotional beings – operate through the same causal chain:

they stir up our emotions which in turn render us cognisant of the things that we value and care about; this supplies motivation for action and triggers the reflexive process of defining the best way to address our multiple and often competing concerns; which creates the need to continuously self-reflect and self-monitor to ensure that our commitments remain fulfilling and viable in the long-term. Thus, having recognised both themselves and others as individuals whose personal well-being depends on the state of things they truly care about, the readers should have no difficulty in following my analysis of ethical consumer concerns and commitments and their role in shaping individuals' identities.

Before I begin to unravel the process of becoming and being an ethical consumer, I want to introduce my research participants and provide a glimpse of their unique characters and life stories via short vignettes.[1] My aim is not only to set the background for the forthcoming analysis and discussion of findings, but also to allow readers an opportunity to see the real people behind the data and acknowledge all those who took part in my research and made the ambitious undertaking of illuminating and reconnecting the inner worlds and outer lives of ethical consumers possible and, I hope, successful.

Vignette 1: Lucy

Lucy is 48 years old. She is university-educated and works in the field of career guidance and occupational therapy. She tends to identify with the middle class, although she has gone through times of financial insecurity in the past and experienced first-hand what living on a tight budget feels like. At present, Lucy lives with her husband who helps with the grocery shopping and is responsible for most of the cooking, a chore she has always considered too burdensome.

Lucy grew up in a suburban area as the youngest of five children. Her family's diet was a mix of convenience foods – "stuff from cans and packets" – eaten during the week and traditional English meals served at weekends, when her working mother had the time to cook from scratch. Lucy's ethical-consumption practices have at their core a belief in the intrinsic value of all living things and are guided by her commitment to promoting the good of nature and animals. This distinct sense of warmth and respect towards all living beings is rooted in Lucy's childhood experiences: raised in a household full of pets, she was closely involved in the care of animals. Lucy's affinity for animals and her growing passion for protecting their wellbeing began to affect her consumer behaviour quite early in life: at the age of twelve, she proclaimed herself a vegetarian and has abstained from eating meat ever since. Three years ago, Lucy took her ethical consumption project to the next level by switching to veganism. Lucy's self-identification as an ethical consumer is underpinned by a deeply cherished commitment to cruelty-free living. At the same time, she admits that more recently she has become somewhat less rigid in pursuing a strict vegan diet, with concerns over health looming large over her eating choices. For Lucy, being ethical in consumption ultimately means "that I thought through it carefully and it fits with my conscience", the principle of not doing harm being the key moral benchmark against which she evaluates the morality of her consumption behaviour.

I explored Lucy's life as an ethical consumer through extensive face-to-face communication that engendered valuable and trusting relationships. This included two accompanied shopping trips, informal discussions about personal life and consumption ethics, and a formal two-and-a-half-hour interview.

Vignette 2: Jason

Jason is a 31-year-old doctoral researcher of Greek origin. As an overseas student with no family in England, he lives alone in rented accommodation and has sole responsibility for managing all household tasks, including shopping and cooking. Jason refused to identify himself in terms of a social class – to him, the category "class" is an alien concept; however, his level of education, professional status, and material circumstances all point towards membership in the middle class.

Jason grew up in a traditional household where the kitchen was considered to be the women's province: his mother was responsible for preparing all family meals, which were invariably wholesome, healthy, and fresh. Surrounded by small-scale holdings and year-round food markets, Jason's family had easy access to locally grown, organic, seasonal produce. Although questions of food ethics have never been explicitly raised in the house, the family's eating habits were imbued with an environmental ethos – unlike most modern-day western consumers, Jason grew up knowing where his food comes from. At present, sensitivity towards food provenance and the origin of the goods he consumes forms the core of Jason's ethical practice and is reflected in his strong preference for organic, seasonal, and local produce, fair-trade coffee, and sustainably sourced ingredients and natural materials. In England, however, Jason's ethical purchasing intentions are hampered by the lack of availability of products with desirable qualities – hence, imported, out-of-season, pre-packaged foodstuffs have become rather common, albeit unwelcome, elements in his eating routine. Jason defines an ethical consumer as "someone who takes into account a range of different issues, such as responsibility". To him, ethical consumerism is "not the action of just eating" but "the whole life attitude", a position which underpins Jason's continuous efforts to contribute to responsible waste management and recycling.

I met Jason on five separate occasions over the course of my research. These included two informal meetings during which we discussed traditional Greek cuisine, Jason's personal taste preferences, and his farming pursuits; two pre-planned grocery shopping trips; and a formal, hour-long interview.

Vignette 3: David

David is 33 years old; at the time of the study he was pursuing a doctoral degree in the field of sustainability. He felt at a loss about how to define his social position – his working-class background is in a stark conflict with his subsequent educational and occupational trajectories, which suggest membership in the middle class. David is in a stable relationship and lives with his girlfriend, a dedicated environmentalist and life-long vegetarian. For them, grocery shopping is a shared activity, but food preparation is almost exclusively David's domain – a passionate and competent cook, he finds the process of cooking enjoyable and rewarding. David is a vegetarian, avid supporter of organic and local farming, and regular buyer of fair trade. He is highly knowledgeable about the specifics of global food production and its environmental and societal repercussions; his consumption choices are well informed and carefully thought through. At the heart of David's ethical practices are concerns over the environment, animal rights, and social justice. His commitment to a meat-free diet has purely environmentalist underpinnings – raised in a milieu where hunting for food was a common practice, he considers killing animals for consumption to be not only morally acceptable, but a natural part of the cycle of life. For David, being an ethical consumer means "making a deliberate, conscious decision to do what you think is good and always in opposition to what you think is bad". At various points in life, his intentions to consume in environmentally and socially responsible ways were constrained by the lack of money, availability, and accessibility of products with ethical qualities, and the wider social context – the factors which, in David's opinion, constitute the biggest obstacles to wide-scale adoption of ethical consumption practices.

David is one of the few participants that I invited to take part in my study having learned about their ethical consumer pursuits. A committed vegetarian with strong environmental values and a long-held interest in sustainability, he seemed – and proved – a valuable informant. David's contribution to my research included three shopping trips to different food retail outlets; numerous informal discussions, face-to-face and via email; and two separate interviews, each lasting more than two hours.

Vignette 4: Darren

Darren is a 36-year-old vegan and animal rights activist. He earns his living through various commercial activities, such as the distribution of herbal teas, but his main focus is on activism and charity work. Darren identifies as "lower class" due to his poor family background, unstable occupational status, and modest income. His level of cultural capital, however, suggests membership in the more privileged segments of the society: he is university-educated, well-read in philosophy and sociology, and has extensive knowledge of a range of social, political, and environmental issues. Previously married, Darren is now divorced and lives on his own. As such, he is solely responsible for all food-related chores, from planning weekly menus and choosing where to shop for ingredients to looking for recipes and cooking meals.

Darren grew up in a single-parent household, where family dinners alternated between freshly cooked traditional African-Caribbean dishes and ready meals, and convenience foods. Meat has always constituted an important ingredient in his family's cooking, and it was not until quite later in life that Darren began to gain awareness of the ethical and moral issues surrounding the production and consumption of animal products. Darren's sense of consumption ethics has evolved progressively over time: starting off by adopting a pescatarian diet, he later transitioned to vegetarianism, before eventually committing himself to a vegan lifestyle. Nowadays, Darren's consumption decisions are guided by his unwavering devotion to cruelty-free living, a moral position which forbids the use and consumption of all products of animal origin. This, however, represents only one dimension of his ethical life project. Darren's moral concerns find expression in a wide range of other-regarding activities: he runs a charitable organisation promoting a cruelty-free lifestyle, organises community events to feed the hungry, and gives public speeches to increase public awareness and understanding of veganism. While Darren is both highly aware and supportive of the organic and fair-trade movements, he can rarely afford to pay the substantial price premium attached to these products. Most recently, however, he has set up a group to grow organic food for the homeless through collective garden-tending effort. To Darren, ethical consumption means "not causing suffering", and it is because he persistently tries to avoid "taking part in animal abuse and suffering" that he self-identifies as an ethical consumer. Darren considers habit or, in his own words, "mental slavery" to be the biggest impediment to people's transition towards alternative – more socially, environmentally, and morally responsible – ways of consumption.

I got to know Darren very well over the course of my study – as a person, moral agent, and consumer. Our rapport developed progressively through open conversations which occurred – some spontaneously, others intentionally – during accompanied shopping trips, visits to Darren's allotment site, and informal meetings, all leading up to a formal interview that lasted two hours.

Vignette 5: Mary

Mary is a 64-year-old retiree with a master's degree in environmental technology. Over the course of her life, she has held several research and teaching positions at various universities. She identifies as middle class, although, like many of my interviewees, she recognises the fuzziness of the concept. Currently, Mary spends her free time volunteering for environmental organisations, such as the Wildlife Trust. She lives with a lodger and a cat, and does all household activities, including shopping and cooking, by herself.

Mary grew up on a farm, in close proximity to nature and wildlife, and became involved in food growing and animal tending at a very young age. Raised on a diet of fresh, wholesome, traditional meals cooked by her mother, she has grown to appreciate healthy, local, sustainably sourced food. For Mary, ethical consumption is a way of addressing environmental, animal welfare, and social justice issues; she actively seeks out and regularly buys organic, locally grown, and fair-trade produce. Although Mary has never been vegan or vegetarian, over the past several years she began to make conscious efforts to reduce her overall meat intake, a deliberate decision informed by her growing awareness of the environmental and moral issues involved in intensive livestock production. She can be described as a vegetarian-oriented omnivore – a person who, whilst not being strictly vegetarian, displays a tendency to prefer vegetarian options over meat-based meals. For Mary, ethical consumption means "being aware of the impact on other people and other living organisms on the planet of the decisions you make and the things you consume, and trying to create the least negative impact as you can – socially, environmentally, ecologically". It is because most of her consumption choices, albeit admittedly not all of them, "are with an awareness and some consideration of this effect on people and the planet" that Mary considers herself an ethical consumer. In her view, lack of clear information and financial constraints constitute some of the strongest barriers to consumer engagement in ethical purchase and consumption behaviour.

I explored Mary's life and food story whilst accompanying her on her weekly grocery shopping trips (three in total) and during a formal two-hour interview.

Vignette 6: Maggi

Maggi is a 62-year-old retired social worker. She is university-educated and self-identifies as middle class. Maggi is divorced, both her children have already moved out of the family home leaving her responsible for managing all consumption-related chores. Like Lucy, she grew up in a house with pets and developed a close emotional affinity for all living beings. At some point during her childhood, Maggi began to ponder over the relationship between human and nature, and whether eating animals was morally justified. These sentiments were rather out of tune with the social environment in which Maggi was being raised, nor did her incipient moral concerns receive sympathy or support from her immediate family. Both Maggi's parents were fond of traditional British cooking and considered meat an essential component of a "proper" meal; as such, considerations of the ethics of meat eating were simply not present on the family's moral agenda. However, as soon as Maggi left home for university, her growing concerns over the morality of food choice began to translate into a concrete behaviour change, leading her to become vegetarian. Although there were certain points in Maggi's life at which she deliberately returned to meat-eating, such as during pregnancy or breastfeeding, she has sustained a sense of a morally responsible food choice throughout her life. Several years ago, she adopted a vegan life-style and has been abstaining from the use and consumption of all animal-derived products ever since.

While Maggi is a regular patron at health and ethical food stores, she does not shy away from conventional supermarkets, where she seeks out organic produce and ethical alternatives to traditional foodstuffs. Maggi's commit-ment to ethical purchasing and consumption is a manifestation of her deep-felt and well-thought-out concerns over environmental degradation, animal neglect, and social justice. For her, an ideal product choice would be cruelty-free, sustainably produced, ethically sourced, and fairly traded, but prohibi-tive prices of ethical goods make certain compromises inevitable. For Maggi, ethical consumption means "not exploiting people, not exploiting animals, that's healthy, that's sustainable, and it's in terms of simple living". She iden-tifies as an ethical consumer because, as she puts it, "I care about what I eat". Expressions such as "I ought to" and "I should" featured prominently in Maggi's account of her ethical consumption practices, indicating a strong sense of responsibility and commitment. Social pressure and poor catering for vegans and vegetarians are major challenges Maggi faces as a dedicated ethical food consumer.

Over the course of the fieldwork, I have developed a very close rapport with Maggi through extensive email communication and face-to-face interac-tion, including three separate visits to various co-housing sites, four accom-panied shopping trips, and a two-hour-long interview.

Vignette 7: Joe

Joe is a 29-year-old political and social activist. Like most of my inter-viewees, he does not have a very clear class identity, but tends to associate himself with the lower middle class. For the past several years, Joe has been working as a call centre agent at a bank; most recently, however, he quit his job to pursue political and social causes. He is now an active Green Party member and intends to run for his local municipal council. Joe lives with an omnivorous friend with whom he shares kitchen space, but each of them follows his own food practices.

Joe was born into a family of dedicated animal rights activists. Raised as a vegan, he has never knowingly eaten meat except after losing a bet to a friend. During his life, he transitioned back and forth between being a vegan and sliding into a vegetarian diet depending on his living condi-tions and surrounding social context. Underlying Joe's commitment to meat-free consumption are moral considerations and environmental con-cerns. His eating practices are an important part of a larger project of ethi-cal living, which involves continuous efforts to reduce his personal carbon footprint, contribute to local economy, fight animal cruelty, and promote human rights. Joe is consciously trying to follow a harm-free lifestyle, but occasionally finds himself falling short of his vegan principles because of social influences, limited product choice, and even temptation. In his view, the power of habit, lack of personal motivation, and poor social support are the main factors that prevent people from shifting towards more ethical consumption styles. Joe defines ethical consumption as "the purchasing and utilisation of foodstuffs where your primary consideration is not taste or nutritional value, but wider moral principles". He feels very strongly about being an ethical consumer, which for him is reflected in the fact that the ethical implications of his consumption decisions are, as he says, "what I think about first".

Throughout the study, Joe has been very generous with his time helping me to explore his personal pathway towards ethical consumerism. In addition to two pre-planned grocery shopping trips and a two-and-a-half-hour interview, we met casually on several occasions and had extensive discussions about consumption ethics. Moreover, Joe sent me excerpts from his personal food diary, detailing his efforts to shop and consume in a more ethical way. This has offered me a unique insight into Joe's most private deliberations about his consumption behaviour and life practices generally, enabling me to produce a deeper and more accurate account of his personal development as an ethical consumer.

Vignette 8: Solveig

Solveig is a 29-year-old doctoral student with no clear-cut class identity. Originally from Germany, she moved to the United Kingdom several years ago where she now lives with her husband. Together, they are responsible for maintaining a vegan household – despite being a meat-eater himself, Solveig's husband is very supportive of his partner's ethical consumption commitments and willing to cook and eat vegan food at home.

Solveig grew up in a household where fresh sourdough bread was baked every morning, and cooking from scratch was a natural part of the family's everyday life. Reflecting the spirit of traditional German cuisine, most festive and everyday meals served at home or social gatherings were meat-based; to this day, the idea of abstinence from animal products seems bizarre to many of the people with whom Solveig grew up. Yet, from a very young age, Solveig began to deliberate upon the morality of killing and exploiting animals for consumption. Like Maggi and Lucy, she grew up with pets, an experience through which she learned to value the life of all sentient creatures. At the age of nine, Solveig made a conscious decision to switch to a meat-free diet thus committing herself to being a vegetarian among carnivores. She has been an on-and-off vegetarian until several years ago, when several factors prompted her to transition to veganism. Underpinning Solveig's consumption behaviour is the moral principle of not doing harm which, in her view, first and foremost entails refraining from the use and consumption of all animal-derived products. Solveig's ethical practices, however, extend beyond a commitment to veganism. Whenever possible, she uses her consumer power to support local farmers and producers. Solveig is a fervent opponent of the "throw-away" society: refusing to live a consumerist life, she buys most of her clothes from charity shops and combats food waste through freeganism. For Solveig, ethical consumption means "do no harm or do as little harm as possible", and her ethical consumer self-image rests upon her continuous efforts to live in consistence with this moral principle. At present, time and convenience are the key factors inhibiting Solveig's ethical consumer intentions.

I accompanied Solveig on three shopping trips, during which we discussed a wide range of consumption-related moral and social issues. I further explored Solveig's beliefs and commitments in a formal interview that lasted two hours.

Vignette 9: Lila

Lila is 34 years old. She is university educated, works as an editor, and pursues a PhD in the field of cultural anthropology. She identifies as middle class, albeit she only became familiar with the concept upon moving to England. Lila and her husband raise two kids, a six-year-old girl and a teenage boy, her husband's son from his first marriage. Even though both men are meat eaters, the household is kept almost entirely vegan, reflecting Lila's life-long abstinence from animal products. Lila's ethical consumption project extends beyond a commitment to animal welfare and accommodates concerns over the environment and human rights. Her food choices tend to be organic and fair trade; wherever possible, she prefers to buy local, seasonal, and unpackaged produce. For Lila, ethical consumption is "mindful consumption". She considers herself to be a mindful consumer, whilst being highly conscious of the inevitable compromises and inconsistencies permeating her consumption behaviour. In her view, habit and lack of reliable information are major factors inhibiting individual and collective progress towards ethical consumerism.

I came to know Lila through one of my research participants, Maggi, with whom we paid several visits to a co-housing site developed by Lila and her husband. A committed vegan of 20 years, dedicated environmentalist and fair-trade supporter, Lila perfectly fit my research focus. She was the only participant with whom we did not go grocery shopping, the reason being that food provisioning in Lila's household is organised in a truly alternative way. Although Lila patronises two independent ethical grocery shops, she does not have an established shopping routine and relies predominantly on alternative ways of food provisioning. Together with her husband, Lila is involved in a buying group that sources organic and fair-trade products from trusted suppliers at wholesale prices. They also subscribe to a vegetable box scheme, which deliveries seasonal organic vegetables and fruits on a biweekly basis. Lila is strongly opposed to the global food business, hence her exceptionally rare visits to mainstream supermarkets and chain stores.

Since shopping in the traditional sense does not hold a prominent place in Lila's consumption routine, and because her visits to conventional shops are infrequent and always spontaneous, I had to find some other way to gain an insight into her purchasing practices. The opportunity presented itself when Lila invited me to join her and her husband in the discussion of their next food order with the buying group. As Lila scrolled through the pages of an online catalogue, she and her husband exchange views on different products, revealing the complex interweaving of family needs and requirements, personal preferences and tastes, ethical considerations and moral concerns. At the end of this process, a list of products was compiled and agreed upon, which I got a copy of. This experience not only allowed me to directly observe Lila make her food purchase decisions, but it also exposed the complex moral reasoning and judgement behind them. I met with Lila on several other occasions, including a co-housing community event, before conducting a formal interview which lasted more than two hours.

Note

1 I protect the identity of the interviewees through the use of pseudonyms, all of which, except one (Lucy), were chosen by the respondents. In four out of nine cases, however, I adhere to the expressed desire of my interviewees to use their real first names (Darren, David, Maggi, and Joe). I give the participants' age at the time of our first meeting.

Reference

Sayer, A. (2011). *Why Things Matter to People*. Cambridge, UK: Cambridge University Press.

6 Becoming an ethical consumer

Moral concerns, emotional commentaries, and reflexive deliberations

This chapter marks the first stop on our tour of the private and social lives of ethical consumers. In it, I bring the abstract – my proposed theoretical account of ethical consumer identity – into relationship with the concrete – the subjective experiences, feelings, and thoughts of self-perceived ethical food consumers – in order to empirically illustrate the inner psychological process via which the ethical consumer identity comes into being. Here the spotlight is turned on one particular participant, Lucy, whose story provides ample insight into the questions this chapter takes up. To this centrepiece, I bring insights drawn from other participant narratives to add further depth, detail, and nuance to the arguments being made. I focus on the very first steps that individuals take on their journey towards becoming ethical consumers – from gaining insight into the ethical problems around consumption, to developing a sense of concern and responsibility for the social, environmental, and moral goods, to adopting alternative modes of consuming in an endeavour to fulfil their moral duties as citizens of the world. Through in-depth analysis of selectively chosen episodes from the lives of yet-to-be ethical consumers, I empirically re-construct the complex and intricate process which allows the ethical consumer identity to emerge, evolve, and materialise in behaviour.

While it is of course true that identity formation is a continuous process that never stops or slows down, it is also true that certain life events and experiences more intensely than others influence the trajectories of people's identities. Consistent with my proposed theoretical account suggesting that the ethical consumer identity develops through a causal chain linking ethical consumer concerns, commitments, and practices, I explore consumer transition towards ethical behaviour as though it occurs in stages, linked to particular psychological processes (discern, deliberate, dedicate) which enable the production of the ethical self out of the consciousness of an ordinary consumer.

First, I enquire into the origins of ethical consumer concerns by examining individuals' subjective experiences, perceptions, and interpretations of objectively existing conditions, events, circumstances, and investigating their role in triggering the affective and cognitive processes leading people to develop ethical sensitivities regarding the use and consumption of goods. As we delve deep into the inner workings of the minds of ethical consumers, the key role of emotions in alerting individuals to concerns over the ethical problems surrounding consumption and supplying motivation to address them through action becomes apparent, as does the central place of cognition and reason in the development of consumer sense of moral responsibility. I apply the idea of internal conversation to explain why and how emotionality interacts with reason in the process of consumer moral conversion and, drawing on empirical findings and insights, demonstrate the indispensable role of reflexivity in enabling agents to evaluate the social world and their subjective relationship to it and define what they care about and how they should live.

Finally, I show that by designating the morality of consumption as their ultimate concern and by committing themselves to more ethical consumption behaviour, individuals decide not only "how to act, but who to be" (Giddens, 1991, p. 81), that is they define themselves – privately, but also inevitably socially as we shall see later on – as ethical consumers. Here again, I draw on participants' self-accounts to provide empirical evidence for a close link between consumers' adoption of ethical practices and their self-image, thus exposing ethical consumption as a domain where the shaping and reshaping of identity takes places. Whilst focusing on the central role of agential subjectivity in the production of consumers' ethical self, I at the same time acknowledge and examine the ways in which structural objectivity feeds into and determines the outcomes of one's reflexive deliberations about what to do and who to become.

On the origins of consumers' moral concerns

It seems intuitively obvious that for anyone concerned with understanding the origins of people's moral beliefs, commitments, and values, the socio-familial contexts in which they grew up should be the first calling point. Thus, having set myself the task of exploring where ethical consumer concerns take their roots, I turned to participants' earliest memories of their relationship with food and the wider cultural, community, and family contexts in which they developed. Indeed, for some of the ethical consumers I interviewed, the family setting played a truly important role in inspiring ethical attitudes. David, for example, links his deep-seated environmental beliefs to the thoroughly politicised atmosphere he was immersed in at home throughout his childhood and adolescence:

> I was raised that way, that was just what I thought to be normal, so when I got to about 17–18, it is a natural thing, it gets included really – if you are worried about politics, if you are worried about social justice, you are automatically worried about the environment. It just seemed a natural choice.

Mary too began to acquire knowledge about political and social issues and gain awareness of their ethical dimensions in the circle of her own immediate family. Her father's outspoken political views and strong commitment to civic engagement and activism laid the foundation for the development of Mary's ethical consciousness, as a citizen and consumer. "He was very left-wing, so we were very involved with community stuff and support for the miners and political stuff, so right from the start I had a certain value system", she explains. In the same vein, Joe recognises the importance of parental influence in determining the ethical standards of his consumption behaviour and setting the moral tone for his future lifestyle: "I was raised on a strict vegan diet, my parents were vegan, my younger siblings were all vegan … that was just the norm for me to be on a vegan diet."

This evidence is consistent with the widely accepted idea that socio-cultural contexts exert powerful influences on individuals' views and behaviours (Bradford, 2012; Matsumoto, 2007). On the whole, of course, the statement that knowledge, values, and attitudes that people acquire over the life course are to a large extent shaped by their social surroundings is hardly debatable. Yet, as I argued repeatedly in the first part of this book, to suggest that individuals' adoption of certain beliefs and positions is a matter of mere socialisation would be to fall into the trap of downward conflation by reducing human agency, subjectivity, and individuality to the effects of social structures. As Bradford (2012, n/p) reminds us, "it would be a mistake to view societies' socialisation processes as monolithic, 100% effective production lines for the creation of perfectly adjusted and socialised individuals ready to take their place in their preplanned niche in the social world". My analysis of participants' pathways towards ethical consumption confirms that the socio-cultural approach only goes so far in explaining the rise of the ethical consumer and ultimately falls short of accounting for the variety of individual-level factors involved in the process of consumer moral conversion. The inadequacy of "the oversocialised conception of man" (Wrong, 1961) becomes increasingly apparent in light of the stories of individuals who refused "to take their place in their preplanned niche" and, driven by subjectively defined concerns and commitments, chose to occupy roles and positions that were often far from acceptable, appropriate, or desirable in their social worlds.

One such story is told by Lucy, this chapter's protagonist. She was raised in an environment where traditional dishes such as roast beef, meat pies, and stews were favoured and frequently served, where meat was considered to be an essential part of the diet, and where vegetarianism was an almost entirely unknown phenomenon. Neither Lucy's immediate family, nor the wider social context in which she grew up, can be considered responsible for the rise of her ethical consciousness in the consumption domain. When asked about the origins of her concerns about the morality of meat-eating, Lucy began to recall her childhood experiences with animals – how she lived in a house full of pets, and how she had always been "soft on animals". However, unlike most children who would be thrilled to have 35 rabbits in the family garage, Lucy remembers being deeply saddened by the sight of caged animals: "I felt trapped and I kind of identified with all these animals that were in cages, and I thought it's wrong." Archer's conception of emotions as commentaries on human concerns offers a useful theoretical lens through which to interpret Lucy's affective response to the animals' caged existence. As we learned earlier, emotions are relational, that is they arise in relation to something, "and that something is our own concerns which make a situation a matter of non-difference to a person" (Archer, 2000, p. 195). Viewed through the prism suggested by Archer, the affective reaction experienced by Lucy – the feelings of sorrow and sympathy for the misery of captive animals – can be construed as an emotional commentary aroused by concerns over animal life and wellbeing. The emotional import of these incipient but fast-growing concerns was sufficiently strong to supply motivation for action. Lucy vividly remembers how she used to fight against perceived animal cruelty: on one occasion, she let her friend's hamster out of its cage; at another time, she opened the door of the birdcage to set her sister's budgies free. Lucy's experience with caged pets was the first in a series of incidents and events that triggered the development of her ethical consumer concerns accompanied by powerful emotions and a genuine urge to act upon them.

Lucy's ethical sensitivities grew progressively more intense as she developed political and social awareness through engaging with protest music and literature, exploring the pacifist ideology, and participating in the anti-nuclear movement. This was also the time when Lucy's socio-cultural environment expanded immensely following her transition from a Christian school to a multi-cultural school attended by children from diverse religious and ethnic backgrounds. Exposure to alternative worldviews prompted Lucy to challenge the traditional cultural framework within which she grew up and, especially, the Christian idea of man's dominion over animals, which contrasts sharply with the deep respect for all forms of life and commitment to non-violence advocated in Hinduism. When Lucy was twelve, a radio interview with Chrissie Hynde – a rock star, vegan, and animal rights activist – introduced Lucy to vegetarianism and its underpinning ethical principles. It was at this point that Lucy decided to stop eating meat, a commitment informed by a strong emotional aversion to animal cruelty and the reflexive work of comparing and contrasting different outlooks, questioning the accepted, and exploring new ways of thinking, acting, and relating to others. Lucy's commitment to ethical consumption gained further strength and significance in light of a school trip to an animal farm where she, then a fourteen-year-old teenager, was deeply distressed by the sight of animals held in confinement. She remembers being "totally freaked out" upon seeing a sow crammed into a small crate, separated from her piglets by metal bars. Triggered by this experience, a wave of intense emotions washed over Lucy: "I was so shocked, it gave me nightmares, it was absolutely appalling" and reinforced her sense of moral responsibility and commitment as a consumer: "I knew then I'd made the right decision." Three decades later, Lucy gained yet another disturbing insight into livestock farming. During a trip to Switzerland, she was deeply upset and moved to compassion for the fate of little calves locked away on a highland dairy farm. Here is how Lucy recollects this distressing incidence:

> This was in a French-speaking area of Switzerland, up in the highlands where they have a lot of cheese and milk and little dairy cattle – all very beautiful, bells around the neck, you know, it's idyllic, Alpine scenery – really, really beautiful. But, unfortunately, every day we'd walk past these calves who'd been separated from their mom, every day, and they were crying, they were

just protesting against that fate, and it upset my so much – I still feel tears when I think about it.

Profoundly touched by this scene, Lucy experienced the same feelings – a bout of sadness and burning desire to release the captive animals – that she felt when she was a five-year-old girl: "and I was thinking looking at them, if I could get in there, unlock it – I would."

As Lucy attests, this experience, wrapped up in intense and painful emotions, has been central to her irrevocable decision to go vegan: "I just couldn't eat the milk or the cheese, I couldn't do it. And I haven't been able to since." While insights drawn from Lucy's narrative clearly suggest a direct relationship between her experiences with suffering animals and the rise of her concerns over consumption ethics, what needs to be explained is how exactly this relationship was established and in what way it aroused and strengthened Lucy's sense of moral responsibility as a consumer. The concept of glimpsed experience can shed light on the origins of Lucy's concerns regarding the morality of consumption and the source of her emotional aversion to animal products. Viewed through the lens provided by Coff (2006), Lucy's encounters with captive animals represent glimpsed experiences via which she gained an unmediated insight into morally disturbing aspects of meat and dairy production. Through these "glimpses", concerns over animal welfare and life were instilled into Lucy's mind; being genuine concerns, they provoked an intense emotional response and strong desire to take concrete action towards alleviating animal suffering – hence her decision to abandon all animal-derived products and embrace a cruelty-free lifestyle. For Lucy, the scenes that she witnessed – pets locked in cages, pigs behind metal bars, calves in chains "ankle deep in their own muck" – became emblematic of animal suffering and forever associated with meat and dairy, which from then on she was never able to eat again.

At the same time, Lucy does not exhibit the same ethical sensitivity with regard to the environmental impacts of consumption. Concerns over the health of the planet remain largely unconnected to Lucy's personal moral matrix, lack of first-hand experience of or insight into food growing being one of the factors inhibiting the urge to "save the environment". It is noteworthy that Lucy's own commentary supports my conjecture: "I suppose if I was a gardener, I'd be more bothered, you know, if I was growing my own food." Further evidence of a direct link between glimpsed experiences, ethical concerns, and moral commitments leading to more responsible consumer lifestyles comes from the accounts of other participants. Consider, for example, how Lila explained to her little daughter why dairy products have been banned from the family menu:

> I used to explain like, you know, this milk was taken from a cow, and there is actually a calf waiting for this milk, and it is not having it because you want to have it – does it look fair to you?

Intuitively, Lila followed Coff's prescription for mobilising a person's sense of food ethics: she created a story to give her daughter an indirect yet revealing insight into milk production and animal suffering associated with it. This glimpsed experience has achieved its intended ethics-inducing effect by emotionally engaging the little girl – "that's really sad", was the child's affective reaction – and making her feel personally responsible for the unjust treatment of animals. To take another example, Darren's sensitivity towards animal cruelty, which has been growing in him since childhood, transformed into a full-blown moral concern through his being subjected to suffering. For Darren, this occurred via a rather peculiar experience – acute pain and discomfort caused by a piece of meat which got stuck between his teeth. As Darren explains, his own sensations sensitised him to the pain of others:

> I don't want to feel pain, I want to feel as less pain as possible – I want to be pain-free, suffering-free. So then you link that to the wider world, to others, you think about people – what about their pain, other people's pain and other animals' pain? And these things link, connect …

Through his personal experience of pain, Darren gained an immediate insight – a glimpse – into the suffering of other sentient begins and came to closely identify with them. This internal emphatic response fed into Darren's on-going deliberations about consumption ethics and gave him the final impulse to turn veganism into a life-long moral commitment: "the pain, and the concern for the animals, and contradictions with other forms of meat-eating in other cultures, you know, all that came together." In this way, Darren came to embrace a moral duty to avoid bringing suffering and harm to those around him, "eliminating the suffering, taking it out of your life, removing it" has become the key guiding principle of his consumption and living practices.

Mary's case provides an example of the same concern-generating mechanism applying to a non-vegetarian person and engendering ethical consumption commitments that relate to the treatment of nature as well as animals. Mary's concerns over planetary wellbeing have been fuelled by stories about environmental disasters reported by newspapers and television news. These written and pictorial accounts created an opportunity for Mary to obtain a glimpsed experience of the devastating effects of human activity on the natural environment. Through these recurrent glimpses, she developed acute environmental concerns and burning desire to address them in her capacity as a consumer: "It made me think what else should I do, and I wanted to do something about environmental issues, it became more my passion." Likewise, Mary's strong preference for humanely produced meat stems from her first-hand experience of the fear and suffering accompanying the slaughter process: "I've been to an abattoir, I know what abattoirs are like. I know that for animals it is an intense period of fear, and I am not entirely happy about that." Moreover, from a very young age Mary has been actively involved in farming and gardening: she gained extensive first-hand experience of the natural world whilst tending plants and animals on her aunt's small holding and a neighbouring farm. Privileged with ample opportunities to experience food production "in the local and in the present", Mary developed a strong sense of consumption ethics. Her growing concerns over environmental sustainability, ecological balance, and biodiversity have eventually materialised in a stable and long-lasting practice of buying organic produce from local farmers and independent green grocers as well as growing her own food. These findings align with past research indicating that direct experiences of the natural world and participation in nature-related activities can play an important role in the formation of individual environmental values. Based on interviews with climate change activists, Hards (2011) concluded that personal involvement in the eco-regulatory practices, such as tending a garden or animals, is a major factor in the development of nature-respecting values. Not only does my study confirm this relationship, but my interpretation of participants' accounts through the lens of glimpsed experience allows us to understand the underlying mechanism that links consumption-related experiences – direct or mediated, comprehensive or merely "glimpsed" – to ethical consumer concerns, commitments, and practices.

Several important conclusions arise from the above analysis. First, it demonstrates the value of the concept of glimpsed experience for illuminating the sources of consumers' ethical sensitivities and explaining the origins of their environmental, social, and moral concerns. My assessment of participants' personal journeys towards ethical consumption supports Coff's idea of glimpsed experiences as a means whereby modern consumers recognise and engage with moral and ethical issues involved in consumption and cultivate a sense of personal responsibility for the wider implications of their everyday choices. Elaborating on Coff's original theory, I suggest that glimpsed experiences engender a profound ethical connection between consumers and the objects of their concerns: by perceptually, cognitively, and emotionally engaging with the morally disturbing situations, conditions, and circumstances of which they caught a glimpse, subjects make them an indelible part of their personal biographies, their own unique stories, and their self-narratives. Through this relationship, mediated by both emotions and reason, consumers develop a sense of moral responsibility for the choices and decisions that directly affect how the objects of their concerns fare and the situations they care about unfold. The key point about glimpsed experiences is that they operate through compression of psychological – emotional and cognitive – distance between individual consumers and the wider consequences of their behaviour, rendering spatial and temporal gaps between production, consumption, and the ethical impacts thereof incapable of preventing the rise of consumer sense of moral responsibility.

This conclusion shines a new light on the significance of physical proximity in fostering human morality, a question on which some strongly contrasting views have been expressed in the past. Bauman (1989, p. 192), for example, insists on the importance of closeness in enabling and encouraging moral conduct:

> Morality seems to conform to the law of optical perspective. It looms large and thick close to the eye. With the growth of distance, responsibility for the other shrivels, moral dimensions of the object blur, till both reach the vanishing point and disappear from view.

Vetlesen (1993), however, challenges the assumption that there is a necessary link between physical proximity and moral behaviour, arguing that proximity makes a moral difference only through interaction with other factors. What is crucial, he argues, is "the capacity for developing empathy with others ... the faculty that underlies and so facilitates the entire series of specific, manifest emotional attitudes and ties to others, such as love, sympathy, compassion, or care" (Vetlesen, 1993, p. 381). Identity theorists align with the view that identification with the pain and suffering of others, referred to as empathetic understanding, is highly consequential for individuals' formation as ethical subjects: "we behave ethically because we can imagine ourselves in others' stories", argues Lawler (2008, p. 24). Likewise, Steedman (1986) contends that identification with others is central to the process of self-production – we forge our identities, she claims, by putting ourselves into other people's stories, interpreting, reinterpreting, and making them a part of our own biography.

My research findings provide relevant empirical evidence to inform this debate. Participants' stories suggest that the sense of care and responsibility for the other, "this building block of all moral behaviour" in the words of Bauman (1989, p. 184), can be triggered in various ways, which do not necessarily meet the criterion of physical closeness. We have seen how, in Darren's case, the moral urge followed the projection of his physical pain onto other sentient beings, "then you link that to the wider world, to others" and appropriation of their suffering, "what about their pain?", and how for Solveig it has arisen through self-identification with animals: "I am only separated by degrees from, say, a pig or a cow ... How can I justify putting other sentient creatures though treatment like this?" The focus on the relational dimension of moral action underscored by both Vetlesen and Lawler is reflected in the way my respondents articulate the sense of moral obligation in consumption via the narratives emphasising commitment to "doing no harm" and "causing no suffering" to others. This clearly reveals meaningful connections between the respondents and the objects of their moral concerns. In the examples above, however, these connections have been fostered through compression of cognitive rather than physical distance, through empathic understanding, imagination, and identification with others – absent or present, distant or close – for whom moral responsibility is assumed and exercised via forms of ethical consumption.

The ethics-inducing effect of emphatic understanding manifests itself clearly in the modern consumer society – it is reflected in the well-documented ineffectiveness of information-based approaches to changing patterns of consumer behaviour (Barnett, Cloke, Clarke, & Malpass, 2005) and the growing popularity of advertising and marketing campaigns featuring images of food producers and growers, depicting scenes of farm life, and revealing workers' day-to-day struggles for livelihood. Through focusing on presumably real people and their stories, these campaigns aim to emotionally engage their audiences – achieve the "gut reaction", in the words of Lucy – and instil concerns over the wider social, environmental, and moral implications of their personal choices into the hearts and minds of ordinary consumers. In light of Coff's theory of food ethics, claims about the potential of labels, packaging, and promotion materials to interpellate a morally responsible consumer take on a new significance. The idea of glimpsed experience goes a long way towards explaining how exactly texts and images imbued with ethical and sentimental meanings and messages work to induce the "globalising reflexivity" and bring distant places and people "into the world of concern (and pocketbooks) of Northern consumers" (Goodman, 2004, p. 893). What the discursive and visual narratives that "veritably shout to consumers about the socio-natural relations under which [the goods] were produced" (Bryant & Goodman, 2004, p. 348) essentially do is provide consumers with a glimpse of moral and ethical issues involved in the manufacturing process. Through these glimpses, consumers become aware of "the world of meaning" (Goodman, Dupuis, & Goodman, 2012, p. 43) behind the products they choose – such compression of the physical and cognitive gaps between consumption and production ends of the supply chain is precisely what renders consumers capable of extending their sense of moral responsibility to distant parts of the world (there are apparent parallels here between Coff's idea of the ethics of distance and what Goodman (2004) calls "political ecological imaginary" with its expansive "spatial dynamics of concern").

Here one may be seduced into thinking that such a scenario suggests a passive – "constructed" and "governed" – consumer whose ethical choices are merely a part of the system of "transnational moral economy" (Goodman, 2004) into which she gets drawn by carefully designed and strategically disseminated discourses of which she is neither a master, nor even a co-producer, but only a slave (I have discussed this position in detail in Chapter 1). This view appears to rest on the assumption that the tools and devices deployed in the project of the governing of consumption possess some kind of self-energising and self-extracting symbolic power which sucks unaware and unreflexive consumers into the global moral economy irrespective of whether, how, and with what outcomes they take up and make sense of the sentiments, ideas, and meanings being thrust upon them. It seems obvious, however, that such ideological propaganda, however skilfully crafted, can only achieve its intended effects in the presence of active consumers, able and willing to engage – through emotions and reason – with textual and visual symbolic narratives presented to them. As Adams and Raisborough (2010, p. 258) note with reference to Newholm (2005):

> Studies of ethical consumption campaigning may well point to the 'generation of narrative frames in which mundane activities like shopping can be re-inscribed as forms of public-minded, citizenly engagement' (Clarke et al. 2007b: 242) but analyses of people's own accounts of their consumption practices suggests (sic) that such re-inscription is not wholly manageable or predictable.

This is because morally charged discourses do not simply produce conscientious consumers as social subjects out of cultural dopes or involuntarily impose the ethical consumer role model on unsuspecting individuals; instead, they "energize consumers to be morally reflexive" (Goodman, 2004, p. 896), that is they engage them in reflexive deliberations about the objective moral duties they have in virtue of being consumers and how they can best fulfil them from their subjective positions and contexts. It is because of the inherently subjective nature of such internal self-dialogues that the outcomes of reflexive self-scrutiny induced by the glimpsed experience will vary between individuals, leading to differences in the content and extent of the ethical commitments that each morally concerned consumer will be willing and able to undertake.

Second, consistent with the account of ethical consumer identity proposed in this book, insights emerging from the data suggest the centrality of emotions to the process of consumer moral conversion. The statement about the importance of the emotional content of glimpsed experiences represents a refinement of Coff's original account. In his work, Coff invites us to think of glimpsed experiences as sources of information and factual knowledge about consumption-related issues which, when made available to consumers, render them cognisant of and, consequently, sensitive to the ethical implications of their own practices and behaviours. The findings that emerged from my study, however, indicate very clearly that glimpsed experiences stimulate not only thinking (*logos*), but also feelings (*pathos*); they not only increase people's knowledge and awareness about the social, moral, and environmental problems around consumption, but they also unleash powerful affective responses which play a vital role in arousing consumers' ethical sensitivity and propelling them into ethical actions. The feelings of sadness and compassion for nature and animals, the mental distress and discomfort caused by the glimpsed experiences, as well as the enthusiasm, desire, and passion for positive change are well captured in the participants' narratives of becoming ethical consumers. The emotional charge of ethics-inducing experiences is reflected in the use of strong emotion phrases and words, such as "shocked", "mortified", "passion", "rush of horror", by the respondents. Products and goods, therefore, become not merely "silent documents", as suggested by Coff, but also strong emotional anchors: their symbolic power extends beyond the ability to remind consumers of the relevant glimpsed experience and includes the potential to revive its accompanying emotional states. Lucy's account of her encounter with calves locked in chains on an Alpine farm tellingly illustrates this relationship. Let us consider the following excerpt again:

> everyday we'd walk past these calves who'd been separated from their mom, every day, and they were crying, they were just protesting against that fate, and it upset me so much – I still feel tears when I think about it.

The above quotation reveals that Lucy's experience was infused with emotion and that those emotions were of a certain kind: she displayed understanding of the animals' misery, empathised with them, and felt deeply saddened by their suffering. This particular incident, as Lucy repeatedly stressed, provided the main impetus for her transition to veganism: "when I came back to England I just thought, oh well, that's it ... I got that feeling – I just can't eat dairy, can't do it". The ethics-inducing effect of this glimpsed experience, however, cannot be attributed to its information content: having been aware for a long time of the ethical issues surrounding dairy, Lucy learned nothing new about the realities of milk production from her daily visits to pasture. Yet, a direct, unmediated encounter with the pain and suffering experienced by the captive animals triggered a powerful emotional reaction which provided a driving force for an immediate change in Lucy's consumption. Admittedly, several other factors played an important role in enabling Lucy to switch to a meat-free diet, increased public awareness and availability of green goods being some of the major societal developments which rendered veganism a feasible undertaking, for Lucy and for many other consumers driven by the same moral principles. However, the critical element that was missing from Lucy's previous attempts at going vegan, and that provided the final impetus for her firm commitment to a vegan lifestyle, was a profound emotional connection with the objects of her moral concerns, forged through the glimpsed experience of their pitiful existence. The following excerpt from my interview with Lucy conveys exactly this point:

Lucy: I've tried being vegan a few times before this time. This time it's worked because of those calves that were looking at me sympathetically. So now it's worked, now I just could not drink milk, I could not do it.
Me: But what was different before?
Lucy: I suppose I hadn't got that emotional revulsion. I've got to have that gut reaction and just think – no.

The "gut reaction" and the "rush of horror" to which Lucy referred again and again when talking about her emotional and physical aversion to animal products are the resounding echoes of her original affective response to the encounters with suffering animals. These unsettling incidents are brought back to Lucy's memory by the mere sight of meat and dairy products, which for her became the "silent documents", the embodiments of the pain, harm, and suffering inflicted on animals: "if someone was offering me a bacon sandwich, I'd immediately think of that". These disturbing memories are not just structured records of Lucy's experiences – they are also depositories of the painful emotions which were aroused by them and whose intensity does not seem to be fading with time: "I still feel tears when I think about it", says Lucy, her facial expression and the tone in her voice attesting that the feelings of sadness and sympathy for the little calves retain their grip on a moral agent inside her.

Lucy's deep commitment to cruelty-free products contrasts with the rather tepid support she lends to environmentally friendly goods – a disparity which begins to make sense when considered through the lens of emotions. In fact, Lucy explicitly refers to emotions, or rather lack thereof, when explaining why she is not committed to buying organic and local produce to the same extent as to being a vegan: "it does not get me emotionally in the same way as, you know, animals", she admits. While Lucy is highly aware of the environmental impacts of consumption and sympathises with the moral causes at the heart of the fair trade, organic, and local food movements, she does not have that all-important emotional connection with them which is needed to supply motivation for action.

> I can see it's important, I am not denying it's important, but, you know, I don't feel a rush of horror from eating something that's from Kenya in the same way as I would over eating sheep's eyeballs.

This example supports the idea that emotions constitute a necessary link in the causal chain leading consumers from merely being aware of a particular ethical issue to wanting to act upon it. Admittedly, certain practical constraints would have to be negotiated had Lucy decided to commit to a more environmentally friendly diet, the higher costs of ethically labelled produce being one obvious hindrance. However, before objective constraints to such an ethical food project could even be reviewed and evaluated, relevant concerns must have been logged into Lucy's moral register and elevated to the rank of ultimate concerns. The lack of emotional response – "a rush of horror" in Lucy's words – to imported foods suggests that environmental issues, such as pollution and climate change, failed to rise to the top of Lucy's subjective hierarchy of concerns.

Joe offers another telling illustration of the ways in which human concerns and emotions link up to incite agential actions. For him, a job in the banking industry became a source of profound personal dissatisfaction: "I became very unhappy with the direction my life was going in." The growing feeling of disaffection experienced by Joe was an emotional commentary upon his deep-seated concern over the right way to live: "I knew in my gut it was – or felt in my gut – that it was wrong working for the bank". This affective reaction to an increasingly unrewarding lifestyle propelled Joe into specific actions: he quit his job at a bank, joined the Green Party, and resumed a vegan diet – all as part of a comprehensive effort to address his nagging concerns over ethical living: "that was the prompt to go back to that trying to be a bit more ethical …". The important role of emotions as a source of the "shoving power", which moves morally concerned consumers to ethical actions, becomes even more apparent in light of the explanations provided by the respondents for their disinclination to engage in certain forms of ethical consumption. The lack of affective response is a key reason why eco-conscious Mary continues to eat meat: "I think for a lot of vegetarians – the ones I came across – it was an emotional response about not killing animals, and I am okay about killing animals, actually"; and why David, another dedicated environmentalist, decided to pursue a vegetarian as opposed to a vegan diet: "I just don't feel bad enough about it, I just don't feel bad about milk and cheese". These examples confirm a direct causal relationship between human concerns, emotions, and actions theorised by many thinkers, including critical realists. Supporting Archer's account of human emotionality, Sayer suggests that "emotions involve desire and concern to produce or prevent change; they incline us to act in some way, though we may override such inclinations" (2011, p. 37). The argument that "it is not possible to have a genuine concern and to do nothing about it" (Archer, 2007, p. 231) gains strength in light of the remarks made by my interviewees: "I care about these things, and I can't care about them, and think about them, and know about them without enacting that" (Mary); "people who say they do care and do nothing – they don't really care, that's not caring. It's got to be linked to action if it's a genuine thing" (Joe); "if you have some ideas you really believe in, it only means something if you manifest these ideas in your life … you have to just walk the road and make it your own, otherwise it's just so superficial" (Lila). These claims should, of course, be considered in the context of an acknowledged need to account for the effects that objective structural factors exert on individuals' ability to embark upon and pursue their preferred courses of action. Chapter 7 provides such an account.

While the concept of glimpsed experience and the idea of emotions as commentaries on human concerns offer a way to understand how ethical consumer concerns emerge and move people to actions, reflexivity and internal conversation cast light on the subjective properties and inner processes which enable ethically sensitised consumers to define their ultimate moral concerns and devise projects through which to address them. In Chapter 3, I have theorised this process and its underlying mechanism; let me now illustrate it by presenting examples from the respondents' self-narratives.

Reflexivity and internal conversation in the context of ethical consumption

My analysis of participants' self-accounts lends support to the view that internal conversation serves as a medium in which the reflexive work leading individuals towards their ultimate concerns and commitments takes place. In the following quote, David shares his inner deliberations through which he came to designate nature as his ultimate concern and embrace environmentalism as his most important personal commitment:

> I decided that there is this kind of hierarchy which is similar to the way that ecological economists think about the Earth as a system. So you have the Universe, and you have the Earth, and inside the Earth you have life, and inside life you have society. If you are going to choose which one to start with, you go as high up the chain as you can, and I decided just because of that simple logic that the thing I am going to care about is the environment.

David's musings epitomise Archer's idea of the internal conversation as reflexive self-talk during which subjects deliberate upon the external world and what matters to them in it, discern and sift through their various concerns, and, finally, dedicate themselves to those they consider to be most important and worthwhile to live by. Another glaring example of an internal conversation comes from Solveig, who engaged in an inner self-dialogue with herself about herself – her relationship to the world and her moral duties as a human being and as a consumer – in an effort to work out the most moral way to act and to be.

As a human being am I really that separate from a chimpanzee? And if I am not, how am I ... I am only separated by degrees from, say, a pig or a cow ... if I don't have to eat these products to survive, how can I justify putting other sentient creatures though treatment like this?

In a similar way, Darren conducted an internal conversation with himself about the practical and moral significance of his various concerns on his way towards veganism: "you have to ask yourself how important it is, those foods – those eggs and that milk – how essential it is for health, can we live without it?" Maggi also engaged in reflexive deliberations when devising her ethical consumption project. Not only did she determine how she felt about consuming in an environmentally responsible way: "I really don't want to buy things that aren't organic", but she also assessed the concomitant costs of such a commitment: "I'd be spending a ridiculous amount of money" and considered it in light of competing concerns: "it's that conflict with how much money I am prepared to spend on food". In these examples, we observe internal conversations whose purpose was to lead consumers towards their ultimate concerns and commitments through understanding their emotional import, moral worth, and the sacrifices involved.

A closer look at the inner reflections shared by the participants leads to an important conclusion, namely that "values are things people can reason about" (Sayer, 2011, p. 18). Armed with this evidence, I want to join critical realists in a crusade against a widely held misconception that judgements of value and objectivity do not mix. My participants' reflexive deliberations about their moral concerns, responsibilities, and ways to fulfil them, demonstrate not only that "emotions and subjectivity influence how we reason and what we accept as fact" (Sayer, 2011, p. 18), but also that reason has a large role to play within emotion and value. This argument is best illustrated by Lucy, whose horror at the sight of pigs locked away inside cramped cages and profound sympathy for them stemmed from her understanding that animals are sentient beings, capable of experiencing emotional and physical pain: "I know pigs are intelligent animals and can be very loving ... and they got this sow, and she'd had a litter, and she was separated from them by iron bars ... that totally freaked me out". These embodied gut feelings immediately fed into Lucy's judgement of the moral relevance and significance of her ethical consumption commitments: "I knew then I'd made the right decision". To take another example, Maggi's moral concerns kept growing in intensity as animal abuse and cruelty in factory farms were becoming increasingly recognised and exposed to the public eye:

That's become more talked about, and that kind of knowledge and information is around more over the years. It's become more important for me, I suppose, with the realisation about animal welfare and that milk and eggs as well were cruel – the whole production process was.

Similarly, Joe's lifelong commitment to a plant-based diet, initially little more than a habitual practice into which he was conditioned by his vegan parents, took on new significance when he became more aware of the global environmental and social impact of the modern economy: "I was like – oh, isn't capitalism terrible, it's destroying the planet and the environment, and such a big thing I can do to combat this would be to go vegan again". His pursuit of a vegan lifestyle became a more purposeful, more consciously lived, and much more reflexive practice when he gained a deeper insight into its fundamental ethical underpinnings:

> I think the real difference was being at a college studying sociology, and I did philosophy and ethics as well. And then living with my aunt who was very much into [veganism] and that was a big part of her identity, and that very much changed my reasons behind it.

Knowledge and reason played an equally important role in Darren's internal conversation about the right way to consume. His reflexive deliberations about the ethics of eating were informed by insights drawn from philosophical teachings: "I read a book … it had a lot of facts and arguments in for vegetarianism … so that was quite enlightening"; science: "there is enough evidence to suggest that we are supposed to be herbivores … medical evidence shows that as well"; and religious texts: "I'd look at things at the Bible, religions, what you can eat and what you can't eat … that informs your ethics, you are trying to make sense out of these different patterns and contradictions". In Mary's case, news about environmental disasters, such as the Torrey Canyon oil spill, and expanding media coverage of the global environmental movement exemplified by Greenpeace and Friends of the Earth prompted a sense of urgency in addressing nature-related problems: "I became a lot more aware of the environmental issues". Mary's growing environmental and social awareness was, in her own words, "part of the gestalt", the sum of all the intellectual and emotional insights that were streaming into her ongoing internal self-dialogue about her moral responsibilities to the surrounding world – nature, people, and animals. To take a more specific example, Mary's strong commitment to choosing fair-trade products and the emotional satisfaction derived from fulfilling this perceived moral duty of hers have a distinct cognitive underpinning, as revealed in the following quote:

> Because I do know – having studied, having read about these issues a lot more – I am aware of just how much the livelihood of people in developing countries is dependent on growing cash crops, and how much fair trade enables them to live better, and it just feels fairer.

The above comments demonstrate that the working out of their ultimate ethical concerns and commitments by morally inspired consumers occurs through the interaction between emotions and reason. This conclusion aligns with the findings from psychological research into the process of consumer moral conversion. In a study on moralisation of eating practices, Rozin, Markwith, and Stoess (1997) distinguish between affective (e.g. first-hand experiences of animal slaughter) and cognitive (exposure to information about animal welfare issues) factors influencing consumers' adoption of ethical lifestyles. Likewise, McDonald's exploration of the psychological process of becoming a vegan draws attention to the cognitive and emotional contents of catalytic experiences through which people "*comprehend*, as well as *feel*, the consequences of the new knowledge of animal abuse" (2000, p. 9, my italics). However, McDonald's account only describes the empirical – the subjective experiences acting as catalysts for respondents' commitment to veganism, and the actual – the consequent changes in their practices and beliefs, whilst staying mute on the real – the underlying causal processes by which the said catalytic experiences bring about attitudinal and behavioural change in ordinary consumers.

Although the important role of both emotional and cognitive factors in stimulating individuals' engagement in ethical consumption has been theorised and confirmed empirically, there is a lack of research shedding light on the mechanism that allows emotionality to come into play with reason to enable ethically minded consumers to work out their ultimate concerns and life commitments. Drawing on Archer, I suggest that this mechanism is an internal conversation during which individuals, through recourse to their reflexive capacities, clarify, organise, and prioritise their concerns, decide what their ultimate life projects are, and how, if at all, they can be realised under the given conditions. Concerns, therefore, constitute the main theme in our inner self-dialogues, and it is because the objects of our concerns exist independently from our being concerned about them (just as emotions emerge in relation to something that is ontologically independent from any human consideration) that our internal conversations are always infused with both emotion and reason. As Archer explains, since our concerns involve "both an external 'object' and a subjective commitment, the outcome itself will be a blend of logos and pathos" (2000, p. 231). In his insightful exploration of why things matter to people, Sayer (2011, p. 1) too contends that concerns "are not just free-floating 'values' or expressions projected onto the world but feelings about various events and circumstances that aren't merely subjective". As he argues emphatically, "keeping feelings separate from thoughts, echoing the body-mind dualism, is absurd; both are responses to the world and our concerns" (Sayer, 2011, p. 39).

In this book, I join the chorus of those insisting on the co-constituting duality of emotion and reason in the process of reflexive scrutiny of the self and its relationship to the world. The results of my own inquiry into the private workings of the minds of morally inspired consumers reveal the key role played by both *logos* and *pathos* in the evolution of ethical-consumer concerns, practices, and identities. They suggest very clearly that participants' decisions to commit to ethical consumption were driven by both their emotional responses to the situations and matters of concern and reasoned judgements about how they can address them through personal projects and actions. It is noteworthy that many of the respondents explicitly acknowledged the important role of reflexivity in enabling individuals to work out the relationship between their inner self and the outer world, set their ultimate life goals, and work their way towards them. Participants' awareness of the centrality of continuous reflexive self-scrutiny to achieving and maintaining an authentic life is revealed in their comments on the mental journey towards ethical consumption:

I think that comes from examining life much more … it's always a good thing to take a look at where you are, build a clear picture of where you are and where you want to be, and then move towards that (Joe);

You are having to look what is the wrong thing to do, what is the right thing to do, in order to do the right thing you need to replace, find another way of living (Darren);

I think it is important to think about what you are doing, you know, not act unreflectively (Lucy).

They agree, therefore, that "self-inquiry ultimately enables an authentic self through active and deliberate choices of a specific, ethical consumption lifestyle" (Cherrier, 2007, p. 322).

Considered through the lens provided by reflexivity and the internal conversation, the inner psychological process, whereby ethically sensitised consumers arrive at their ultimate moral concerns and consumption commitments, becomes more transparent. It remains to be seen, however, whether and how the reflexive process of working out what to care about affects what kind of persons agents become and how they view and relate to themselves. In the remainder of this chapter, I present my analysis of the meaning and relevance of participants' ethical consumption commitments for their sense of self-identity. I thus aim to show that the close-knit relationship between ethical consumer concerns, practices, and identities theorised in this book holds up empirically.

Consumption and the ethical self: on the links between concerns, practices, and identities

The main reason I am vegetarian is just because … It is such an ingrained part of my identity … It's like the fact that I am male, you know, that's my identity now, I feel it's so solid and unchanging.

(Joe, ethical consumer)

This comment from Joe reveals the intimate connection between ethical consumption and self-identity. Like most of my participants, Joe could not be more explicit and emphatic about the deep significance that his commitment to a morally responsible lifestyle bears for his sense of authenticity and being true to himself. Unprompted, he suggested that ethical consumer identities are ultimately grounded not in specific behaviours and practices, for those are merely external manifestations of our internal commitments, but in their underlying moral concerns – things that people truly care about, that they see as their most important mission in life, and that they are willing to dedicate themselves to, despite the sacrifices and challenges such commitments involve. In the following quote, Joe conveys this idea through a simple, yet revealing and rich metaphor:

> Being vegetarian or being vegan is ultimately just a dietary thing, so if you switch from eating foods from all colours to just eating blue foods or not eating any foods that have blue – that does not change you. It's the things that drive you to be vegetarian or vegan – that's where the change comes from.

To further illustrate this important point, let me consider the explanation provided by Lucy as to why she so strictly avoids products from factory farms. Her answer is resolute and heartfelt: "it is just not right, I can't do that to animals, it's just cruel", she says unwaveringly. Upon closer scrutiny, Lucy's statement conveys two distinct, but equally important ideas, namely that she considers factory farming to be a source of animal cruelty, and that she does not want – cannot bring herself – to participate in it. Through the juxtaposition of the two points: "I can't do that" and "it's just cruel", Lucy draws a sharp demarcation line between herself and the act of cruelty. The reason she refuses to be implicated in animal abuse is not simply because she considers it cruel, but because she does not think of herself as the kind of person who is capable of such cruelty. What Lucy implies, but fails to vocalise is that she is not a cruel person, and it is precisely for this reason that she "can't do that to animals".

More empirical evidence of direct links between ethical consumer concerns, practices, and identities can be found in the comments provided by other participants. Echoing Joe, Darren describes his transition to ethical eating as a process of finding his authentic self in the things that he sincerely values and cares about: "it was like the search for the truth I guess, in learning who you are, discovering who you are, and what is important to you." Similar sentiments permeate Maggi's explanation of why her commitment to cruelty-free living has intensified over the years: "as I get older that becomes more of a focus for me about what's important to me, about who I am". The symbolic, meaning-generating, and identity-defining potential of ethical consumption was asserted by David, for whom "fair trade – organic thing became automatically attached to political notions of environmentalism and social justice"; Lila, who treats veganism as "a way to communicate your ideas and ideology and identity"; and Mary, for whom being an ethical consumer "is kind of part of who I am really".

Furthermore, the belief that ethical consumer concerns constitute a defining feature of a person's identity, to a large extent shapes the impressions that participants form about other people. Maggi, for example, rejects the idea of developing a relationship with a meat-eating person: "it's about who they are", she explained, "if they think it's alright to eat meat then that's part of their value system that would clash, you know".

For Joe, commitment to ethical eating represents an essential aspect of not only his own personality, but also that of his future life partner: "I've not had a long-term relationship with someone who hasn't switched to being at least vegetarian." It seems, therefore, that my respondents concur with the idea that our concerns and commitments are "both extensions and expressions of ourselves" (Archer, 2000, p. 79), and that by choosing what to care about in life one inevitably defines what kind of person she is.

Yet, the argument about the ability of individuals to reflexively define their personal concerns, commitments, and identities should be considered with an important caveat about the limits of agential reflexivity. The attention that Archer's social theory consistently draws to the causal efficacy of human agency should not distract us from the distinctive emphasis that it puts on the role of objective reality in shaping people's identities: "we do not ever make our personal identities under the circumstances of our choosing, since our embeddedness in nature, practice and society is part of what being human means" (Archer, 2000, p. 249). There is a continuous interplay between agential subjectivity and structural objectivity which informs our internal conversations and determines what we commit ourselves to, how we go about our precious life projects, and who we become in the process. This claim is supported by the personal narratives shared by my participants, who displayed unambiguous appreciation of a wide range of influences shaping and moulding their identities. Consider, for example, how Darren celebrates his agential capacity to act independently and make his own free choices: "it was freedom", he says, referring to his unequivocal decision to stop consuming animal products, "it was a choice, it was an ethical food choice, food empowerment and more education, more enlightenment". At the same time, however, he recognises that various social factors, such as the family context, have played an important role in structuring his personal choices and directing his course of action: "your personal family circumstances, depending on how strong their values are, is always going to influence your behaviour". Noteworthy is David's interpretation of his ethical consumption practices as the wilful construction of a desired identity: "it was a lot of just ticking the box – oh yes, I am someone who eats organic, I am someone who buys fair trade, I care, you know, I care about these things, because that is who I am". Like Darren, he has a strong sense of himself as a thinking, intentional agent who actively and reflexively determines which goals to pursue, guided by his own unique system of significance created by means of subjective deliberation upon the objective world: "I decided, I chose to place the environment at the top of my constructed idea of what is important, my hierarchy of what is important." Yet, he too is highly aware that his subjective conception of "what it is right to be" has been shaped by the totality of his personal and social experiences – everything that he encountered or experienced when engaging in interactions with the inner self and the outer world: "all of my experiences had resulted in this my identity". In a similar way, Lucy acknowledges that her distinct personal identity bears the indelible imprints of her social context, present and past: "I am a product of my own environment, of my own background, and my own experiences".

Yet, the fact that ethical consumers may be acutely aware of the objective underpinnings of their subjectively defined concerns and commitments does not diminish or negate their role in the construction of their identities. As Archer (2000, p. 241) argues:

> It does not matter in the least that these concerns do indeed originate outside ourselves in our ineluctable relationship with the natural, practical and social orders, for in dedication we have taken responsibility for them and

made them our own. We have constituted ourselves by identifying the self as the being-with-these-concerns.

Archer's point is echoed remarkably closely by David's insightful and highly reflexive remark: "so just because I know that these things are constructed does not mean I can't enjoy them, allow them to give me purpose, allow them to – allow my identity to develop within them". This is precisely where human reflexivity plays its part as the mediatory mechanism between agential subjectivity and structural objectivity. While not granting agents full control over the properties of the social contexts in which they live and act, reflexivity enables individuals to deliberate upon their subjective experiences of objective reality, identify matters of concern in the events and circumstances that are external to and independent from their reflexive considerations, and in dedication take responsibility for them and make them their own.

In this chapter, I began to trace the consumer journey towards the ethical self. Lucy's story, fleshed out by insights gleaned from other participant narratives, allowed me to illustrate and explain the very first steps involved in the process of consumer moral conversion: gaining a glimpse of the ethical issues around consumption; engaging with the glimpsed experience both cognitively and emotionally – comprehending as well as feeling its wider moral, social, and environmental consequences; becoming concerned about these issues, developing a sense of personal responsibility for how they unfold; being moved by these feelings and, in response to a strong inner urge for action, devising one's own personal project of social change through more responsible modes of consumption. By empirically reconstructing this complex multi-step inner process, I sought to provide convincing evidence for the theoretical model of ethical consumer identity developed and outlined in the first part of the book.

In explaining how concerns over consumption ethics emerge, evolve, and move people to action, I repeatedly drew the readers' attention to the indispensable role of reflexivity in the production of consumers' ethical self. The respondents' narratives of their mental and emotional journeys towards ethical consumption reveal how reflexivity nurtures and drives the internal conversation, an inner process in which *logos* and *pathos* unite to enable ethically sensitised individuals to define their ultimate moral concerns and devise appropriate ways of consuming. Further, insights captured in participants' self-accounts highlight the inextricable connections between ethical consumer concerns, practices, and identities, thus providing support to the claim that in defining what to care about in life (and what they are willing to give up and forego for the sake of their precious life projects) people also define their unique personal identities. This claim, I hastened to add, needs to be read alongside the explicit acknowledgement of the limits and constraints that social, cultural, economic, political, and structural forces place upon agential pursuits of a desired identity. I thus argue that the ethical consumer identity is brought into being and shaped by continuous interactions between social and personal contributions, yet with individual consumers playing an active, highly agentic role in those interactions and taking responsibility for how they unfold and play out in one's relationships with the self and the world. In the next chapter, I will focus closely on the incessant interplay between agential subjectivity and structural objectivity in ethical consumption and the important ways in which it shapes and reshapes ethical consumer practices and identities.

References

Adams, M., & Raisborough, J. (2010). Making a difference: ethical consumption and the everyday. *The British Journal of Sociology*, *61*(2), 256–274.

Archer, M. (2000). *Being Human: The Problem of Agency*. Cambridge, UK: Cambridge University Press.

Barnett, C., Cloke, P., Clarke, N., & Malpass, A. (2005). Consuming ethics: articulating the subjects and spaces of ethical consumption. *Antipode*, *37*(1), 23–45.

Bauman, Z. (1989). *Modernity and the Holocaust*. Ithaca, NY: Cornell University Press.

Bradford, S. (2012). *Sociology, Youth and Youth Work Practice*. Basingstoke, UK: Palgrave Macmillan.

Bryant, R., & Goodman, M. (2004). Consuming narratives: the political ecology of 'alternative' consumption. *Transactions of the Institute of British Geographers*, *29*(3), 344–366.

Cherrier, H. (2007). Ethical consumption practices: co-production of self-expression and social recognition. *Journal of Consumer Behaviour*, *6*(5), 321–335.

Coff, C. (2006). *The Taste for Ethics: An Ethic of Food Consumption*. Dordrecht, the Netherlands: Springer.

Giddens, A. (1991). *Modernity and Self-identity*. Stanford, CA: Stanford University Press.

Goodman, D., DuPuis, E., & Goodman, M. (2012). *Alternative Food Networks*. London, UK: Routledge.

Goodman, M. (2004). Reading fair trade: political ecological imaginary and the moral economy of fair trade foods. *Political Geography*, *23*(7), 891–915.

Hards, S. (2011). Social practice and the evolution of personal environmental values. *Environmental Values*, *20*(1), 23–42.

Lawler, S. (2008). *Identity*. Cambridge, UK: Polity Press.

Matsumoto, D. (2007). Culture, context, and behavior. *Journal of Personality*, *75*(6), 1285–1320.

McDonald, B. (2000). "Once you know something, you can't not know it": an empirical look at becoming vegan. *Society & Animals*, *8*(1), 1–23.

Newholm, T. (2005). Case studying ethical consumers' projects and strategies. In R. Harrison, T. Newholm & D. Shaw (Eds.), *The Ethical Consumer* (pp. 107–124). London, UK: Sage Publications.

Rozin, P., Markwith, M., & Stoess, C. (1997). Moralization and becoming a vegetarian: the transformation of preferences into values and the recruitment of disgust. *Psychological Science*, *8*(2), 67–73.

Sayer, A. (2011). *Why Things Matter to People*. Cambridge, UK: Cambridge University Press.

Steedman, C. (1986). *Landscape for a Good Woman*. London, UK: Virago.

Vetlesen, A. J. (1993). Why does proximity make a moral difference? Coming to terms with lessons learned from the Holocaust. *Praxis International*, *12*(4), 371–386.

Wrong, D. H. (1961). The oversocialized conception of man in modern sociology. *American Sociological Review*, *26*(2), 183–193.

7   Being an ethical consumer

Exercising moral agency in the contexts of objective reality

> The relation that holds between us and our contexts is always one of forces in tension, in which we push and pull, and are pushed and pulled.
>
> (Sayer, 2011, p. 104)

In the preceding chapter, we saw nine unique individuals set out on their personal journeys towards the ethical consumer identity. We heard them voice their ultimate moral concerns, their most intimate feelings, and their privately held reflexive self-dialogues, thereby gaining insight into the inner psychological process by which an ethical consumer is formed and the emotional and mental work that accompanies it. We have seen how, having figured out their subjective relationship to the world as one of concern over the morality of consumption, our subjects committed themselves to alternative – environmentally, socially, and morally responsible – ways of consuming, thus attaining the desired ethical consumer identity. At this point, the focus of my investigation shifts from the process of becoming an ethical consumer to the intricacies and complexities of being one. In this chapter, I proceed with my inquiry into the lives of nine morally concerned individuals and continue to unravel their personal stories to explore how, having defined themselves as ethical consumers, my respondents actively sustain their desired identities through the continuous reflexive monitoring of their ethical consumption commitments against dynamically changing subjective contexts and objective states of affairs. My aim here is to illustrate the complex interplay between structure and agency and their respective contributions to ethical consumer practices and identities, thereby revealing that the ethical consumption phenomenon is shaped by a wide range of factors operating at different levels. This specific objective is part of a larger undertaking the urgency and significance of which I have argued for throughout Chapter 2. It is my hope that by exposing ethical consumption as a site of on-going interactions between agential subjectivity and structural objectivity, this book will begin to correct the imbalances underlying the prevalent understandings of consumer behaviour and promote recognition of the complex ensemble of individual and systemic powers which motivate, inform, and define (ethical) consumption.

Contextualising consumer practices

> There are a lot of structural elements that get in the way of you achieving what it is right to be.
>
> (David, ethical consumer)

The contextual embeddedness of ethical consumption becomes manifest as soon as a person makes his or her first attempt at engaging in alternative ways of consuming. The properties of the natural, practical, and social contexts in which agents are placed, and the relationships which agents develop with them, play a key role in determining whether, to what extent, and at what cost consumers will be able to act upon their moral concerns and take up the desired ethical positions. A review of participants' experiences as aspiring ethical food consumers is helpful for illustrating this argument.

Take Lucy for example. Her endeavours to transition to veganism span more than two decades – the possibility of enacting her ideal vision of ethical eating has, for a long time, been precluded by the practical difficulty of sustaining a plant-based diet in an overwhelmingly meat-eating environment: "this is back in the 1990, and it was not easy, and the sort of food that you got in health food shops was pretty horrible", she recollects. It is revealing how Lucy's commentary on the ways in which her opportunities to make ethical choices changed depending on the social and cultural contexts in which she found herself at different points in life: "it was just really hard, particularly in Moscow, there was nothing in the shops apart from bread and jam"; "Sophia was easy for food, really easy, the fruit and veg at the market were lovely"; "I didn't eat so well in Paris because the monks did not really understand vegetarianism." Thus, for a long time Lucy's repeated attempts to abandon all animal products have been futile – unavailability of vegan options rendered her desired ethical project unrealistic, food being an absolute necessity of life. Presently, when lack of suitable options or unresponsive social contexts no longer present insurmountable barriers to Lucy's pursuit of an ethical diet, her ability to exercise her consumer agency continues to depend on the practical opportunities to do so. The following quote reflects her frustration:

> What annoys me is unclear labelling. I wish they'd labelled things more clearly in situations where I am eating out, like through work when they have a buffet or something and they don't label things, and they don't tell me what I can eat, what I can't eat.

The above comment highlights the important role of systemic measures, structural provisions, and practical devices, such as product labelling, in turning consumer "oughts into cans" (Barnett, Cloke, Clarke, & Malpass, 2005, p. 31). Other participants provide more examples of a key role of contextual factors in determining the ability of individuals to engage in ethical consumption and the ease with which they can do so. For instance, Joe described how his well-established practice of ethical shopping was thrown into confusion upon moving to a new city, where the absence of fresh food markets meant that more of his grocery shopping had to be done in supermarkets and chain stores: "it was very convenient just to nip across to a Tesco and get a sandwich for lunch ... I was a bit unfamiliar with everything, I just found it hard to carry that on [shopping ethically]".

Equally elucidating is Manasi's comparison of the extent of practical effort and commitment that she felt was required to sustain a plant-based diet in the "meat and potatoes land" of Midwestern America versus vegetarian-oriented India: "over here [America] it feels like you have to seek out vegetarian food options, over there [India] you have to seek out meat". Her other remark reiterates my argument about the contextual dependency of ethical consumption: "knowing your farmer is a wonderful thing if you are lucky enough to live in a place where you can do that". These sentiments were shared by Lila, whose experiences as a socially situated consuming agent are remarkably similar to those of Manasi. Lila's allegiance to veganism was easily accommodated in her native Israel, where vegan options are well integrated into the local food and socio-cultural landscape: "most common street food that you get in Israel is vegan: its either hummus or falafel in pita bread, and that's vegan, so that was ok and just completely normative". Sustaining a vegan diet proved equally straightforward at home, where Jewish food traditions were strictly observed:

> Because it is a Jewish family and they keep kosher, so it means they don't mix milky stuff and meat stuff ... it made it really easy for me to become vegan because of this reason, because it was really easy to avoid milk.

In the United Kingdom, however, certain forms of ethical consumption, such as eating local, became more challenging to fulfil due to structural limitations: "some things you just can't buy in the local shops", justifies Lila her involuntary, yet unavoidable visits to supermarkets. Her strong preference for consuming locally grown produce is further constrained by the climatic conditions in which her ethical consumption project unfolds. "It would be easier to go local if you lived in a normal climate ... In England it just means that in some season you just eat kale all the time, and I am not willing to go there", says Lila, offering an example of the embeddedness of ethical consumption in the natural order of reality. Further, the success of David's pursuit of goods with desired ethical qualities has also been always contingent upon his immediate context. Back in his native Scotland, the lack of shops selling environmentally friendly produce was a major constraint to David's pursuit of ethical living: "we had to go to Glasgow to get different things, but you can't go and get your weekly shopping in an hour bus drive away", he explains. This practical barrier itself stemmed from the specific socio-cultural environment that prevailed in David's hometown: "it does not have a very much diversity of people there, so even if you opened a shop selling different things, there were not many customers for it". David's opportunities for making ethical choices have expanded once Waitrose became part of his local shopping scene:

So many times over the years I have been buying things, something I really liked and I felt bad about it, and I thought to myself – I wish I could get the fair trade version of it, I wish there were an organic one of these, and then going to Waitrose and there was.

Another glaring example of the role of surrounding contexts in facilitating as well as restraining consumer engagement in ethical practices is offered by Solveig. Contextual factors proved critical to Solveig's ability to initiate and sustain her ethical food project. She came to the idea of veganism at the time when pursuing a plant-based diet was becoming increasingly easier due to the continuously growing availability of alternative products, including dairy-free and meat-free foods. This important enablement was conditioned by the Green Party's rise to power and the concomitant increase in environmental awareness among the mass public in her native Germany: "I think Green Party government raised a lot of awareness for ethical food production and consumption, so the variety of food offered everywhere and just the consciousness and awareness of people changed", she explains. The key factor that enabled Solveig to stay loyal to a vegetarian diet at various points in life comes down to accommodating social contexts, as revealed in the comment she made about her experience of being a vegetarian student: "it was fairly easy to stay vegetarian, the university cafeterias all had vegetarian options". Unprompted, Solveig attributed the ease with which she was able to engage in ethical consumption in England to the high levels of environmental awareness prevailing in the country and its historically conditioned tradition of accommodating people of diverse cultural backgrounds:

In England, it's a lot easier to be vegetarian or vegan. I think it's because you have a lot of people of Indian and Pakistani heritage, so a lot of supermarkets offer really broad variety of fancy vegetables and legumes and stuff. And also because you get Indian and Pakistani restaurants which often do vegan options or vegetarian options.

Most recently, sustaining a vegan diet became more straightforward for Solveig due to the rise of the Internet, social media platforms, and online communities, on which she relies for information: "god bless the Internet – I would have died without having access to vegan recipes"; news: "when Oreos turned vegan I found that on one of these groups"; and knowledge sharing and support: "sometimes just giving people tips – there is Leeds vegan group, for example" In the context of the key argument of this chapter, noteworthy is Solveig's acknowledgement of her privileged position as a member of an advanced western society and her appreciation of the inaccessibility of ethical lifestyles to the people whose objective conditions are different from her own:

For me it is an ethical obligation not to harm where I don't have to harm … but, of course, that is me, because I live in a western society where I can just go to a supermarket and buy fresh food and stuff everywhere.

However, Solveig accentuated the structural limits of her consumer agency when describing how the absence of fresh food markets close to home makes shopping at conventional supermarkets a more frequent activity than she would have preferred; how UK supermarkets' security measures prevent her from dumpster diving, which she used to practice in Germany; and how her aspirations as an ethical consumer are constrained by the forces of global capitalism: "I would like to consume more products from smaller independent companies, but it is really tricky because you have three or four really big companies that produce soya products and it is very hard to avoid that". Like Solveig, Jason takes up a distinctly anti-corporate, anti-capitalist position, and like Solveig, he is prevented from acting upon his ideas by the properties of the social order in which he lives and to which he – through a conscious choice – conforms: "that is the system, I have to follow it, I make most of my shopping at supermarkets".

On the whole, there was a clear realisation among participants that their opportunities for making ethical food choices are contained within the actual political, economic, and business realities. This was especially evident in the comments capturing the challenges of accommodating diverse ethical concerns in one shopping basket: "I remember Morrisons used to do bananas – you could get fair-trade bananas and you could get organic bananas, but you could not get together, and I remember thinking – should I get fair trade, should I get organic?", recounted Maggi. Similar tensions in attempting to exercise moral agency within the commercial realities of the global food industry were highlighted by Lila: "Do you support a chain and get your fair-trade bananas or do you want to just support your local shops and get those other bananas which may not be fair?" and Joe: "Am I letting down the local business or am I exploiting foreign farmers through using the local business?" Here I would be remiss not to reiterate the inadequacy of choice-theorists' framing of consumer behaviour founded on the assumption of rational self-interest. As participants' comments make clear, ethical choices often involve handling difficult moral dilemmas rooted in concerns for the other – a predicament that can only arise before normative, morally conscious agents who are part of an inter-dependent, inter-subjective reality and bearers of relational, other-regarding reflexivity.

A more detailed enquiry into the role of social contexts and interactions in shaping the respondents' projects of ethical consumption reveals that consumers' ability to engage in ethical behaviours and the ease with which they can do so are strongly affected by the social groups, networks, and institutional relations of which they are a part. Lila, for instance, highlighted the enabling effect of a "lefty" work environment in which she could freely envision and enact an alternative consumer subjectivity: "it was a fairly progressive environment and people were very open to different ideas". In the same way, Solveig commented on her social milieu and its liberating influence on the ethical consumer within her: "because it is a very diverse crowd of people anyway, I have never had any problems". Conversely, engaging in desired behaviours becomes much more challenging when it requires transgression of the particular ways of thinking and acting dominating within a given social context. Joe attests to the truth of this argument:

"it's very easy to be vegan, if you are hanging around all of those animal rightsy people. But when you are out of that sphere, I think it becomes much harder". This view is grounded in his personal experience as an ethically concerned and committed consumer: for him, it became incomparably easier, both practically and psychologically, to engage in ethical food practices when he started to live with his aunt – a devoted vegan, a nutrition professional, and a skilful, competent cook. The following comment captures these important social influences on Joe's pursuit of veganism:

> Living with her, being exposed to all kind of literature in her house, her being so knowledgeable about diet and where to get the right things to eat, and advocating for it ... It was easy to do living with my aunt, very very straightforward.

David's experience of becoming a proponent and practitioner of alternative forms of consumption is remarkably similar to that of the subjects whose views we have just heard. In his case, a more active and explicit pursuit of an ethical lifestyle was rendered possible by specific social enablements, namely a falling out with his old group of friends that David described as "tough, dangerous, bad people" and to whom he could never reveal his ethical self out of fear of being ridiculed, despised, and rejected, and a romantic relationship with an environmentally oriented girl who became "an enabler" of his ethical consumer identity:

> All the things that I wanted to change about myself, I wanted to try – I was only able to try that because of her, because she allowed that, she would not make fun of it, because she was interested in this as well.

These findings align with the literature emphasising the socio-cultural embeddedness of ethical consumption. In a study exploring individuals' adoption of environmental beliefs, Hards (2011) describes how climate change activists deliberately retract from their green identities when those clash with the social world. Her analysis highlights the need to reconcile "competing demands of 'normality' and 'sustainability'" (Hards, 2011, p. 37) as a common challenge faced by environmentally concerned people. The examples quoted above offer further support to Hards' (2011, p. 33) conclusion that "without conducive social networks it may be hard to reject dominant norms, or envision alternative forms of normality" (for an extended discussion of the role of supportive social networks in individuals' participation in alternative consumption movements see also Cherry, 2006).

Overall, participants' experiences demonstrate that consumer choice is characterised by ubiquitous contextual embeddedness, and that multiple systemic factors affect individuals' motivation and potential to be ethical in consumption. More broadly, they suggest that objective contexts in which agents are placed to a large extent determine whether and at what cost they will be able to pursue the desired consumption behaviours. These findings severely undermine the assumption of agential capacity to freely choose and reflexively (or rationally) appropriate regardless of the wider cultural, economic, and political contexts. My analysis empirically demonstrates the frailty of the agency-focused framework by highlighting how respondents' ability to engage in ethical consumption and actualise their desired identities is contingent upon the specific structural conditions that continuously shape their situations and opportunities. It, therefore, provides support to those arguing for the need to avoid "over-exaggerating the reflexive and self-conscious sensibilities" (Adams & Raisborough, 2010, p. 256) of consuming agents and "take into consideration the context of context" (Askegaard & Linnet, 2011, p. 381) in examining consumption processes and phenomena. More generally, it speaks against the existing tendency to neglect the social embeddedness of human reflexivity and overemphasise the freedom of choice at the expense of acknowledging the role of social structure in shaping the self and its practices.

Reinstating consumer agency

Having demonstrated the contextual embeddedness of ethical consumption and the limits of agential freedom to make morally desirable choices, I turn to the other side of the agency-structure equation. Here I provide empirical evidence attesting to the capacity of consuming agents to reflexively monitor their ethical commitments against continuous changes in their subjective concerns and social contexts, to reconsider their ultimate priorities and ways of acting upon them in light of the constantly evolving knowledge of their inner self and the outer world, and to actively negotiate the objective enablements and constraints to their preferred life and identity pathways. In doing so, I will show that the causal powers of social structures do not operate unmediated and unconstrained, and that agential responses to structural properties – both enabling and constraining – play a key role in shaping ethical consumer practices and identities. Together with the preceding section, this discussion will provide an empirical account of "how involuntary placement in the three different orders [of reality] intertwines with the voluntary human response" (Archer, 2000, p. 249) in the context of ethical consumption.

As we have seen earlier in this chapter, unaccommodating and unresponsive social contexts can present a significant impediment to consumers' pursuit of ethical practices. Thus, adopting and sustaining non-mainstream eating behaviours in traditional food environments requires the exercise of agency and ability to question and transgress the dominant social order. Such have been the experiences of Lucy and Solveig, whose agential powers enabled them to commit to what was considered a radically alternative diet within the conventional meat-eating familial and socio-cultural contexts. Like other participants, they manifest the capacity to uphold their consumption commitments despite objective barriers, such as lack of options with desirable ethical qualities. Solveig, for example, negotiates meat-focused events by bringing her own food: "I would bring vegan burgers or sausages so that I would have something to put on the barbecue." The same approach has been adopted by Lucy, who succeeds in staying social without sacrificing her vegan lifestyle: "usually if I go out on a social occasion I take something with me that I can eat"; and Lila, who maintained a habit of bringing her own food to dine on with colleagues during night shifts at work: "I kind of coped, I brought my own packed dinner with me".

Lila's case provides a telling illustration of the role of consumer agency in creating and shaping ethical consumption. In an effort to opt out of the supermarket practice of packing products "with three layers of nylon and plastic", Lila joined a community group that buys foodstuffs in bulk: "I don't feel so guilty about all this packaging because I have just one big 5kg bag of something, I don't have to buy a new lentil bag every month". Taking her environmental concerns further still, she has actively participated in defining the packaging practices of her food suppliers: "we changed our farmers several times ... it was like, can you just pack it with a little less plastic, and can we return the boxes, and can you reuse them ...".

Joe offers yet another example of the causal efficacy of consuming agents. As an undergraduate student, he was faced with the need to defend his commitment to veganism against the lack of meatless options at the university canteen: "I remember having to fight for that for a bit, for that special treatment". Mary adds more evidence to support the view of consumers as active agents willing and able to negotiate structural barriers and constraints. In the following comment she describes how, upon moving to Scotland to take up a new job, she started to actively shape her food provisioning practices thus demonstrating the capacity to take responsibility in consumption and enact change in her immediate food environment:

> I found it pretty limited what you could buy, I couldn't get the food I would have normally eaten ... I set up and started digging out a vegetable bed at the back of a tiny unit that we were working in.

To take another example, prior research has highlighted the higher costs of ethical products as a key factor undermining consumers' ability to exercise moral choice (McEachern, Warnaby, Carrigan, & Szmigin, 2010; PriceWaterhouseCoopers, 2008; Sudbury & Böltner, 2010). Adams and Raisborough's (2010) study revealed that for many ethically conscious consumers, shopping is a "balancing act between [their] social conscience and the size of [their] purse". Tuning in with this sentiment, my respondents provided many examples of the ways in which concerns over limited budgets interfered with their commitments to ethical shopping. The following comments reveal how at various points in life participants had to suspend or abandon their ethical projects in the face of financial difficulties:

> I could not afford to be fussy about whether something was organic or anything like that (Lucy);

> I think I ought to buy organic apples and I did for a few months, but they are twice as expensive (Maggi);

> We had very little money as well, so I could not invest in buying those expensive products (Lila);

I could not find any vegan burgers which were not pre-fried tofu, which was just so expensive (Joe).

Overall, therefore, respondents agreed that ethical consumption involves constant compromises between ethics and costs. This was most clearly articulated by Mary: "if I didn't have enough money, then I wouldn't be able to [shop ethically] and that's that". However, while the hefty price tags of ethical products do indeed place constraints on the contents of participants' shopping baskets, they actively seek out opportunities to engage in ethical consumption in the ways that do not command a premium price. For example, Joe continuously experiments with vegan recipes in search of the most cost-effective weekly menu; Darren set up an allotment group to grow organic food for personal consumption and charity; Lila joined a buying collective to purchase fair trade and organic foodstuffs in bulk at a more affordable price; and Solveig engaged in freeganism in a bid to cut her grocery costs while simultaneously responding to the problem of food waste. Those who for various reasons, such as convenience or lack of alternatives, do most of their grocery shopping in supermarkets, demonstrate equal resourcefulness in finding ways to exercise moral agency at no extra cost. Maggi ensures an on-going supply of ethical products by seeking out special offers and deals – once a bargain is found, she places a bulk order that usually lasts until the next promotion is offered in-store. David remains a regular patron of the upscale Waitrose, where the reduced section is his constant source of otherwise unaffordable goods. "Waitrose is not expensive, you can have expensive things if you want them, or – not", he says, revealing the potential for active choice and resourceful approach to food provisioning. These examples showcase how through creativity and skilful use of available resources, be it money, a plot of land, or a supermarket dustbin, individuals manage to push the boundaries of their living contexts and engage more fully and fruitfully in more ethical modes of consuming. This evidence underwrites the realist assumption of an inherently fluid, transformable reality which changes in response to agents' continuous attempts to adapt the surrounding environment to their concerns, desires, and needs. On larger temporal and spatial scales, this is reflected in the progressive expansion of ethical goods into mainstream food outlets and increased social awareness of and accommodation to ethical consumption, which over the past several decades has moved from the fringes of consumer society to its very core in some contexts.

The above analysis shines a new light on the links between ethical consumption and economic capital. The view that responsible consumers are almost exclusively high economic capital consumers permeates public and media discourse on ethical consumption, and different commentators have described the growing trend towards consuming ethical products as "yet another way in which the poor are being disenfranchised" (Worth, 2006, n/p) and "just another way of showing how rich you are" (Monbiot, 2007, n/p). The tendency to portray ethically conscious consumption as an exclusive province of the rich stems from the fact that ethical goods, such as organic and fair trade, tend to come with significant price premiums, suggesting that the vast shoes of a responsible consumer can only be filled by high-income earners. This is by no means an empty claim – in fact, a comparative assessment of 75 products at the top six UK grocery stores revealed that on average ethical goods are 45% more expensive than conventional products, and that nearly half of British consumers are unable or unwilling to pay the price (PriceWaterhouseCoopers, 2008).

Yet, there are many good reasons for resisting the tendency to equate the ability and willingness to engage in ethical consumption with the amount of economic capital at one's disposal. As Littler (2011) points out, being rich does not necessarily lead to being concerned about the ethical impacts of one's own consumption and lifestyle choices. Indeed, research shows that ethical consumer behaviour is unrelated to income or occupational status (Laroche, Bergeron, & Barbaro-Forleo, 2001; Loureiro, McCluskey, & Mittelhammer, 2002; Pepper, Jackson, & Uzzell, 2009) and that educational level and cultural capital are much better predictors of consumers' willingness to commit to alternative ways of consuming and bear associated costs (Eurobarometer, 2011). In fact, there are many less well-off individuals among those who adopt pro-social and pro-environmental behaviours. Consumer activism among working-class people has a long history, well exemplified by the British co-operative movement and the 1966 boycotts of supermarkets by American housewives (Littler, 2011). Moreover, contemporary statistical evidence suggests that ethically minded consumers are increasingly found among the poorer nations. In a study of the impact of the global recession on the ethical goods market, Carrigan and De Pelsmacker (2009, p. 681) draw attention to the findings of the Havas report indicating that in China, Brazil, Mexico, and India more consumers are prepared to pay premium prices for green products than in the United Kingdom, United States, or Germany. It has also been noted that while certified organic and fair-trade products often come at a higher cost, plenty of ethical choices require no extra spending and allow a convenient marriage of consumers' environmental and financial concerns (Flatters & Willmott, 2009).

The empirical insights into ethical consumer behaviour presented in this chapter lend support to those arguing against the representation of ethical consumption as exclusively a wealthy shopper's pursuit. In fact, only one among my respondents, Mary, was sufficiently well-off to afford buying high-end produce at artisanal farmers markets and specialty food shops. All other participants could not escape being driven by financial considerations when making consumption decisions, and some had to carefully plan their daily and weekly menus to be able to meet monthly bills. Consequently, shopping for expensive fair trade and organic products was by no means the only or most popular way in which my respondents expressed their ethical self. Instead, they engaged in behaviours that required little or no extra spending commitments and even yielded some savings – earlier in this section, I provided examples illustrating various approaches and strategies which participants resort to in an attempt to reconcile their material and moral concerns. Thus, while there is no denying the fact that some people face fewer constraints in their pursuit of ethical lifestyles than others, since, as Barnett and colleagues rightfully point out, "the material and socio-cultural resources required for engaging in self-consciously ethical consumption are differentially available" (2005, p. 41), the claim about the elitist and exclusive nature of ethical consumption overlooks the vast array of consumer practices and activities that fall under the term. The inadequacy of representing ethical consumer behaviour as a privilege only the moneyed class can enjoy becomes apparent as soon as one moves away from the narrow interpretation of ethical consumption in terms of leisurely shopping for conspicuously labelled, premium-priced products at high-end markets and stores. To account for the variety of ways in which people provide goods and in which they may do so more ethically, we need to take the phenomenon of ethical consumption outside the boundaries of artisanal markets, specialised aisles, and quirky shops, and consider a much wider range of contexts in which it belongs.

Overall, the evidence of consumers' ability to actively pursue and promote desired forms of consumption undermines the idea of the socially constructed and governed consumer, for the manner in which participants overcome objective constraints to their ethical food commitments presupposes human agency and capacity to evaluate and respond to social structure. Through their varied enactments of ethically conscious consumption, my respondents affirm themselves as "pluralistic, heterogeneous, and multiskilled ethical persons" (Cherrier, 2007, p. 322), and as agents whose consumption decisions are clearly more than "inculcated responses explicable only by reference to more objective social forces" (Soper, 2007, p. 217). This means that agential reflexivity regains its place in consumption activities, for it is the powers of reflexivity that enable consumers to deliberate upon objective reality and their subjective relationship to it and in light of the reflexively achieved insight devise fulfilling and sustainable ways of addressing their moral concerns. The role of agential reflexivity in shaping ethical consumer intentions and actions will become apparent once we take a close look at how my respondents negotiate their relationship with the natural, practical, and social orders of reality to sustain a satisfying and liveable balance between their multiple competing concerns.

The many concerns of a single (ethical) consumer: striking the balance

As evidenced by participants' descriptions of their internally conceived, yet externally conditioned enactments of ethically conscious consumption, multiple structural and personal factors affect people's purchasing and consumption behaviours throughout their lives and may foster or hamper consumers' success in maintaining ethical lifestyles. The continuity of ethical consumer commitments is contingent upon the ability of consuming agents to uphold a liveable balance between their ineluctable concerns, some of which may be in a fierce competition. To achieve the desired state of equilibrium, consumers will have to continually monitor, reassess, and fine-tune their relationships with all three realms of the world – natural, practical, and social – whilst staying acutely attuned to how this affects the unfolding of their ethical consumption projects. The complex interactions between objective reality and consumer subjectivity are well captured in participants' rich and detailed descriptions of their constant endeavours to consume in a more socially, environmentally, and morally responsible way. To illustrate more vividly these interactions and their influences on ethical consumer commitments, I will consider how Lucy's project of ethical eating fared at different times and under different circumstances.

At first, Lucy met with success in implementing her decision to become vegetarian: her mother was tolerant of her daughter's convictions, and Lucy was allowed the freedom to choose what to eat at home and what to pack for her school lunches. However, as a fourteen-year-old, Lucy, being the only vegetarian in a carnivorous household, was subsisting mostly on vegetables and bread, complemented with sweets and low-nutrient snacks from the family's kitchen cupboard. Soon, Lucy's new eating habits began to produce undesired effects on her health, thereby throwing out of order her otherwise harmonious relationship with the natural realm. After several months on the "bread and jam diet", Lucy fell severely ill. She was diagnosed with anaemia and strongly advised to switch to a more nutritious eating regime. Determined to stay true to her ethical consumption commitment, Lucy refused to go back to eating meat – instead, she decided to address her pressing health problems by adopting a more balanced vegetarian diet. To facilitate her daughter' endeavours, Lucy's mother bought a vegetarian cookbook and allowed Lucy to prepare her own meals – an arrangement which, as Lucy notes in hindsight, involved a significant compromise on the part of a devoted housewife who considered the kitchen her own province.

Let me now analyse the meaning and significance of this episode in the context of Lucy's formation as an ethical consumer. To begin with, I construe Lucy's illness as a physical manifestation, a tangible outcome of the first clash between her commitment to ethical eating and other concerns, which she arose simply by virtue of her being human and which she could not prevent from emerging. In the case at hand, these were health-related concerns, juxtaposed against concerns over the ethics of eating meat, gave rise to what Janda and Trocchia (2001, p. 1208) describe as the "animal-welfare versus self-welfare tension". Once allowed to emerge, this tension could be neither ignored nor willed away: with increasing intensity, its accompanying emotional commentary was calling for immediate attention and relevant action, demanding that the harmful relationship between Lucy's body and her newly adopted diet be improved or eliminated. A difficult dilemma arose before Lucy: she could either abandon her vegetarian lifestyle – this, of course, would involve a departure from her ultimate concerns and, by implication, her authentic identity – or somehow mitigate the unintended effects of her ethical food project on her body and health. As the story reveals, Lucy managed to remedy her relationship with the natural realm by taking advantage of the specific practical and social enablements that were available to her at the time. On the one hand, mastering the art of vegetarian cooking was a performative achievement that enabled Lucy to improve her nutrition and attend to her physical needs without receding from the position of an ethical consumer. On the other hand, Lucy's shift to a healthier diet was greatly facilitated by an important social enablement – the support and cooperation offered by her mother. Thus, by fine-tuning her relationships with the natural, social, and practical orders of reality Lucy managed to restore a liveable balance between her different concerns and preserve her commitment to ethical eating.

Lucy's battle for the ethical consumer identity does not finish here, however. Some years later the lack of dietary variety and overconsumption of nutrient-poor convenience foods led to a new, even more serious conflict between Lucy's ethics and her bodily needs. A doctor's recommendation to start eating fish as a way to increase the intake of protein was initially rejected by Lucy on ethical grounds. The unfavourable status quo lasted until the adverse health effects of poor nutrition and its intensifying emotional import could no longer be repressed or neglected by Lucy. A decisive change in her ethical practices occurred during a trip to Portugal, where a lack of catering for vegans and vegetarians presented a serious constraint to Lucy's commitment to a plant-based diet. An acute allergic reaction triggered by the overconsumption of eggs (omelettes being almost the only vegetarian meal served in Portuguese eateries), easy access to fresh seafood, and words of encouragement from her pescatarian boyfriend, prompted Lucy to reconsider her food ways. Here is how she describes the turning point:

> I got really spotty, and we were in a restaurant – it was a nice outdoor restaurant – and they were doing grilled sardines, you know, that traditional. And he was sat there, and he was eating these sardines, and he said, "they are absolutely delicious, you should eat this" … So I said, "ok, I'll try one", and I ate a sardine. Since then I try and eat fish once a week, just for the health.

Once again, the sum total of influences arising from the natural (health problems), practical (unavailability of meat-free options), and social (the presence of her pescatarian boyfriend) orders of reality led Lucy to revisit her commitment to a strict vegetarian diet, consider it in light of her other pressing concerns, and adjust her *modus vivendi* to reflect the unwelcome but inevitable change in priorities. Having been rather dismissive of the need to maintain a well-nurtured body in her younger years, Lucy eventually had to acknowledge the ineluctability of health-related concerns and adopt a less stringent approach to consumption. Examples from other participants' narratives further highlight the multiple ways in which competing concerns affect consumers' willingness and ability to persist in their ethical projects. Solveig, for instance, confessed to the fact that her health and wellbeing take priority over her ethical views in the situations that require a definitive choice: "if it is something you need or something you don't have an alternative for, then I have to say that my own life in this situation is for me more important than my convictions." In the comment below, she describes one such compromise necessitated by her constant battle with diabetes:

> I went over to a friend's house and it was at night, so there were no shops open or anything. I could feel that I am starting to get a little bit dizzy, and he

said, "well, basically the only thing I have that will give you a quick sugar fix is chocolate". And I thought, okay, I can now basically collapse on the floor or eat chocolate.

Lila went through a similar experience of having her body – its cravings and needs – compete with her ethical principles. The following quote captures the deliberate mental and emotional effort whereby Lila took control over an uncomfortable bodily state:

> I was really starving, it was late at night and we couldn't find anywhere to eat and the only open option was basically McDonald's ... and I said – no way, I am not buying anything from this place ... I better be hungry.

In this case, Lila refused to surrender consumption ethics to her body's immediate urges. Through intentional, active choice symptomatic of agency, she subdued natural desires and instincts prompting her to deviate from commitment to ethical eating. Lila's example supports the idea that once we embark on a particular moral project, "we are no longer capable of the simplicity of purely first-order response: reactions to relevant events are emotionally transmuted by our ultimate concerns" (Archer, 2000, p. 242). It provides a telling illustration of the process of emotional elaboration, whereby individuals actively engage with their emotions and elaborate them beyond primitive reactions and spontaneous impulses so as to protect their most precious commitments. This again highlights the key role of reflexivity in the process of becoming and being an ethical consumer, for it is our reflexive powers that fuel "our ability to reflect upon our emotionality itself, to transform it and consequently to reorder priorities within our emotional sets" (Archer, 2000, p. 222). Which of the competing concerns will be assigned the ultimate priority depends on an individual's subjective judgement of their relative importance and worth, as well as their concomitant sacrifices, losses, and costs. Like other respondents, Lila agreed that in some circumstances she should be prepared to suspend her ethical principles, as, indeed, she had done in the past. Her experience of being both a young mother and devoted ethical consumer demonstrates how concerns about the morality of consumption can be overridden by the pressing demands of life:

> I really made a point of not buying any packed fruit and vegetables, but if you go to a supermarket, it is so much easier just to grab a bag of carrots rather than pick them individually ... And it sounds really lazy, but actually, you know, when you have this big trolley and the girl is whining, you just kind of grab and just go.

Mary gives us another example of how consumers may be forced to put their ethics on hold when the exigencies of family life demand so. During the course of my research, Mary was the main carer for an older relative – every couple of months, she would travel to the other end of the country to look after her aunt who was suffering from Alzheimer's. It was during those travels that Mary consciously set aside her commitment to ethical eating and, having neither time nor the right state of mind to seek out products with desirable qualities, filled her shopping basket with less ethical but more readily available options. Like other participants, Mary interprets her involuntary compromises in terms of a temporary shift of moral focus and mental and physical energy towards those concerns which appear to be more important or urgent at a given point in life and, therefore, need to be put before concerns over food ethics: "I can't worry about it because I've got so many other pressures on me, it is just how it is. When I come back, I'll start again", she explained. Her comment reveals that, like all agents in pursuit of their precious life projects, morally committed consumers are burdened with the need to maintain equilibrium between their natural, practical, and social concerns. Darren's brief yet insightful remark captures the point: "you have to balance enjoyment, your lifestyle, and your culture".

Socio-cultural norms too have the potential to induce compromises in ethical practices. Concerns about meeting social expectations around particular situations and roles play a non-trivial role in guiding ethical consumers' choices and behaviour, as revealed by the following comments from David, who puts aside his ethical principles to be a considerate guest: "To go somewhere else, to stay at someone else's house – you're not going to say, oh, is this avocado organic? I'm not going to eat it then, that's so ridiculous", or thoughtful host: "My guests will get whatever they want, whatever they need and there will be no questions as to whether it is bad for the environment or not, because they are my guests". How morally driven consumers arbitrate between their social and ethical concerns as well as how such arbitration affects their identities and ways of expressing them is another key theme, which the next chapter will explore in detail. The above examples are only meant to give the readers a glimpse of how the inherent human sociality and its ensuing concerns affect the extent and continuity of ethical consumer commitments.

The analysis presented here demonstrates that individuals' perseverance in relation to ethical consumption depends upon their ability to sustain a healthy relationship with each of the three worldly realms and successfully resolve the inevitable conflicts between them. This finding supports the claim that in considering ethical consumer practices and behaviour we need to take account of the ways in which "the moral complexities of everyday life restrict the adoption of an active consumerist role" (Jacobsen & Dulsrud, p. 2007, p. 469). This level of insight, however, cannot be attained without understanding how consumers recognise, evaluate, and respond to concerns that arise on their ways and compete for their attention, and how they conceive and achieve the right balance between them.

Reflexivity in ethical consumption: an ongoing imperative

In this section, I analyse participants' feelings and thoughts about the ongoing changes in their ethical consumption projects and practices. My aim is to illustrate the inner workings of the mind of an ethically concerned consumer engaged in a continuous reflexive monitoring of her subjective moral commitments, the objective contexts in which they unfold, and a constant dialectic between them. The following comment from Mary is a clear example of such reflexive musings:

> I have noticed that there is much more of world food cooking going ... and I started to think – well, my diet shifted that way and I am eating a lot more imported foods and not as much basic English food. And I am thinking – this is going to be affecting world food trade, and people in developing countries, and food growth patterns, and climate change, and all sorts of things. I am thinking – I might have a look at that in my own diet, think about that a bit.

This quote demonstrates how Mary's approach to ethical consumption is underpinned by relational reflexivity operating within an inter-subjective, interdependent social reality. It suggests not only that she stays alert to the ways in which changing economic and socio-cultural landscapes affect her diet, but that she also repeatedly re-assesses the ethical implications of her consumption decisions and continuously reviews the consistency between her moral principles and her eating habits. In the same vein, Lila describes how her food choices change along with the changes in the spatial and informational contexts of her ethical commitments:

> For about a decade I refused to have processed food in my house, and then I read something about super-ethical company that is the most ethical company in Europe ... and I looked at the ingredients and it looked fine, and I thought – my kids are going to be delighted with this processed soya sausages.

Like Mary's, Lila's food practices undergo incessant transformation "because the situation changes as well and I learn more things all the time". Manasi offers another example of an incessantly evolving project of ethical eating:

> You have to change with the times and you have to change with the environment around you ... when my parents were growing up nobody knew what was going on the farms, nobody knew how many pesticides were being used, there was no information, but now that there is information, you can make better choices.

Not only do these accounts reveal the reflexive effort involved in ethical consumption, but they also underscore its continuous nature, problematised by some authors. Thompson and Coskuner-Balli (2007), for example, maintain that a bulk of ethical consumer choices, far from being an outcome of reflexive deliberations, result from consumers' use of heuristics, such as opting for ethical brands and labels which provide mental shortcuts to better purchase decisions. The "ideological allure of simple choices", these authors argue, steers consumers away from a reflexive approach to navigating the complexity of ethical consumption and makes them rely on the simplifying search strategies to achieve the feelings of "confidence in outcomes, direct participatory involvement, and personal engagement"

(Thompson & Coskuner-Balli, 2007, p. 150). Adams and Raisborough's (2010, p. 265) assessment of the moral discourse around fair trade echoes the argument: "the common cultural equation of Fairtrade with 'doing good' might suspend the requirement for reflexive effort otherwise involved in negotiating through the complex demands noted above". While it is hardly contestable that people tend to develop routines for maintaining what they have adopted as their preferred lifestyles, participants' accounts suggest that continuous reflexive monitoring of one's behaviour and actions is an integral part of ethical consumption. Once a person makes ethical consumption his most important life project, his internal conversation becomes "a ceaseless discussion about the satisfaction of [his] ultimate concerns and a monitoring of the self and its commitments in relation to the commentaries received" (Archer, 2000, p. 195). At this point,

> What the subject is doing is conducting an endless assessment of whether what it once devoted itself to as its ultimate concern(s) is still worthy of this devotion, and whether the price which was once paid for subordinating and accommodating other concerns is still one with which the subject can live.
>
> (Archer, 2000, p. 297)

The constant adjustments and changes in participants' ethical practices indicate unmistakably that they engage in such inner work. The most telling illustration is offered by Lucy, who in recent years has become increasingly more flexible about her dietary commitments:

> I am not as stubborn as I used to be, I used to be really really really stubborn and, you know, I made myself ill many times by sticking strictly to a veggie diet. I am not as pig-headed as that anymore. ... I am older, I don't want to be ill, life is short, you know.

Lucy's preparedness to compromise her ethical principles in the ways she would have resented only a few years ago suggests that her subjective concerns undergo constant reordering and review. Having realised – through reflexive deliberation – that she was no longer willing or able to bear the health costs of her dietary commitment, Lucy decided to loosen her grip on consumption ethics thereby tilting the balance in favour of her bodily needs. Lucy's case demonstrates that the ongoing shifts in ethical consumer intentions and practices are informed by the reflexive work of continually evaluating and re-evaluating one's moral commitments and their concomitant costs. To take another example, Maggi engaged in a major review of her ethical practices once her children grew up and left home. In the shift towards increased independence and a reduced level of responsibility for her children's wellbeing, Maggi discerned opportunities to further develop and refine her ethical consumption project. In light of a reflexively achieved understanding of her new circumstances and freedoms, she decided to fully commit herself to a vegan lifestyle. Here is how she explains this long-sought transition:

As I get older, it becomes more of a focus for me – about what's important to me, about who I am … I think because when you're caught up with work and focused on children and making sure that they are kind of healthy and that is the main concern, and rushing around and trying to do everything … and I suppose now that it's just me at home – cause I am not working – and I've got more freedom and space to … focus on the things that are important for me.

The fact that ethical consumption occupied different positions within the set of Maggi's moral responsibilities depending on which concerns were assigned first priority at that particular stage in her life, indicates that a reflexive monitoring of the self and its commitments has been taking place all along, its most recent achievement being Maggi's decision to promote food ethics to the very top of her subjective hierarchy of concerns.

A general conclusion emerges from the preceding analysis. As old concerns go away and new ones arise, as some of them grow in intensity and importance while others become less and less relevant in the changing contexts of life, ethical consumers actively and reflexively revisit and, wherever necessary or appropriate, redesign their ethical consumption projects. The evidence of consumers' ability and propensity to reflexively review their moral commitments and modify their behaviours accordingly belies both the choice-theorists' model of a rationality-driven agent whose ways are pre-defined by a set of fixed and unchanging interests, and the image of the passive, unreflexive consumer whose consumption is governed solely by objective social forces. The reflexive capacity to continuously monitor and revise their practices is indispensable to ethically conscious consumers, because, being a moral project, ethical consumption is not liable to normative routinisation: "since the aim is to determine upon the course of the right action, then 'good' is always the enemy of 'best' (Archer, 2007, p. 301). Participants' attitudes align with this point, as best revealed in the comments from Maggi: "I think it is probably an on-going kind of struggle … struggle of what's best" and David: "the idea of what you think is right to be is constantly moving, constantly changing layer upon layer upon layer". These remarks resonate very closely with the sentiments of fair-trade consumers interviewed by Adams and Raisborough's (2010) who, like many of my respondents, also feel that their ethical activity is "increasingly complicated over the years and requires a constant review and reappraisal of [their] attitudes and values" (2010, p. 262), and that being an ethical consumer requires you to "question your every action" (p. 264), and to be "ethically effective by thinking of the bigger picture *at all times*" (p. 262, my emphasis).

Crucially, however, it is not that ethical consumption is completely immune to routinisation, but that such routines, when and if allowed to form, are constantly challenged and disturbed by the on-going changes in objective reality and subjective priorities. On the one hand, external reality with its multitude of open and mutable systems (Bhaskar, 2013) exerts direct causal effects on agential intentions and actions. On the other hand, the life cycle itself is a source of multiple changes, and, as people make important transitions in their lives – leaving parental home, getting married or divorced, making career progress or retiring, becoming a parent or having children leave the family home – the nature of their relationship with the natural, practical, and social realms change and so do their concerns, priorities, desires, and motivations to pursue certain courses of action (Archer, 2007). It is only by maintaining reflexive awareness of the ongoing flux of personal and structural changes as they occur that ethical consumers can ensure that their moral commitments remain both objectively feasible and subjectively satisfying. At the same time, I would be remiss not to report the limitations of consumer reflexivity, exemplified by the following quotes from my respondents:

> I forget that I am supposed to like this and not that. (David, highlighting the mental strain of having to constantly monitor the fit between one's practices and beliefs)

> It is just neglect. (Lila, explaining the delay in joining an organic box scheme)

> If I was really really committed, I'd spend a lot more time thinking it through and shopping carefully. (Lucy, admitting to the limits of her reflexive endeavours)

These comments reveal that a proportion of ethical consumer choices and decisions remains "outside the area of reflective action" (Jacobsen & Dulsrud, 2007, p. 477) and that mindful consumption is not "a full-time preoccupation" (Jacobsen & Dulsrud, 2007, p. 477) even for the most deeply concerned individuals. Given the acknowledged limitations, the idea of "particular and partial reflexivity" put forward by Adams and Raisborough (2008) may offer a more appropriate conceptual lens through which to consider ethical consumer behaviour. Just as the recognition of the key role of consuming agents in creating and shaping the ethical consumption phenomenon must be accompanied by an explicit acknowledgement of their contextual embeddedness, so should the celebration of agential reflexivity leave room for habitual practices, spontaneous choices, and mindless decision-making in ethical consumption.

Maintaining ethical consumer identities: in pursuit of an authentic self

In the preceding sections, my focus has been on analysing participants' accounts of their personal experiences of being ethical consumers. I have examined how morally committed consumers, burdened with multiple competing concerns, sustain and promote their ethical projects by continuously negotiating a wide range of objective enablements and constraints, and explored the role of agential reflexivity in allowing ethical consumption commitments to persist and flourish through time. In this final section of the chapter, I demonstrate how the continuity of ethical consumption commitments relates to the sense of self, thus providing further evidence of close links between ethical consumer practices and identities.

One significant finding that emerged from this chapter's analysis concerned the unevenness of participants' practices of ethical consumption, a conclusion supported by prior research into ethical consumer behaviour (Adams & Raisborough, 2010). Here I argue that whether or not consumers succeed in fulfilling their perceived ethical responsibilities directly impacts their sense of authenticity, personal integrity, and self-worth, and that the attainment of ethical consumer identity, therefore, is not a self-sustaining achievement, but one that needs to be repeatedly reaffirmed and maintained.

Participants' narratives provide ample evidence of a direct link between ethical compromises and disturbances in one's sense of self. One vivid example is given by Lila, who felt deeply alienated from her true self when she had to abandon, albeit temporarily, her commitment to ethical consumption: "not only I felt guilty, I also felt I was completely remote from myself ... I felt like, who is this person who goes to the supermarket and buys all this packed food? I really felt like it wasn't me". In the following comment, she offers a concise yet profound description of her distressing internal state: "I felt that I did not have the network and the know-how of being really me". The profound identity effects of contradictions exhibited by ethical consumers are further highlighted by Lucy, whose sense of personal integrity was profoundly shaken when she recognised the discrepancy between her beliefs and her consumption behaviour: "I was about 20 when I met vegans and realised that I was a total hypocrite, and there was me eating all this stuff that I should not be eating". Joe feels very akin to Lucy when considering the identity implications of his ethical eating practices: "If I switched from being an ethical consumer of food, I'd feel really hypocritical about that and quite miserable." These comments suggest that for the respondents, performances of ethical practices are not merely acts of consumption, but enactments of their unique personal identities – their true, authentic self. The successes which they achieve and the failures which they commit in their capacity as ethical consumers directly affect participants' sense of personal integrity and self-worth – being deeply invested in their ethical projects, subjects cannot be indifferent to how those unfold. Frankfurt's (1988, p. 83) point is highly relevant here:

> A person who cares about something is, as it were, invested in it. He identifies himself with what he cares about in the sense that he makes himself vulnerable to losses and susceptible to benefits depending upon whether what he cares about is diminished or enhanced.

It is because our concerns and commitments are extensions and expressions of ourselves that participants feel fragmented and inauthentic ("who is this person?", Lila was asking herself) when enacting ethical compromises, and it is for the very same reason that they tend to abstain from making fragile commitments. Here I draw on Sayer's insightful point that "our vulnerability is as important as our capacities; indeed the two sides are closely related, for vulnerability can prompt us to act or fail to act, and both can be risky" (2011, p. 5). Consider, for example, the following comment which explains why, during most of her life, Lucy was reluctant to adopt a vegan diet: "I admired it, I thought it was admirable, I just ...

I didn't want to set myself up and fail." From this statement, we learn that Lucy's unwillingness to undertake a commitment the success of which was uncertain was rooted in fear of failure, one which would not only unsettle her most important life project, but would thereby greatly diminish her sense of self-worth. Again, Frankfurt (1988, p. 83) casts light on the intricate web of connections between concerns, commitments, and self:

> With respect to those we love and with respect to our ideals, we are liable to be bound by necessities which have less to do with our adherence to the principles of morality than with integrity and consistency of a more personal kind. These necessities constrain us from betraying the things we care about most, and with which, accordingly, we are most closely identified. In a sense which a strictly ethical analysis cannot make clear, what they keep us from violating are not our duties and obligations but ourselves.

Consumers' deliberate efforts to ensure continuity and stability of their ethical practices is, therefore, part of their on-going struggle for a coherent, continuous self. The same conclusion has been drawn by Greenebaum (2012) in a study of ethical vegans interpreting consumers' "quest for purity" (p. 131) of their ethical practices as "a pursuit of an authentic identity" (p. 131). This conjecture finds strong support in the narratives told by my study subjects. For example, David's transition to a vegetarian diet was accompanied – and in many ways driven – by the gradual recognition of a misfit between his omnivorous lifestyle and his inner self, defined by concerns over environmental sustainability: "there was this cumulative effect over my entire adult life of this idea that I'm kind of supposed to be vegetarian, but I'm rebelling against myself". A desire to achieve a coherent self-concept by bridging the increasingly obvious gap between his concerns and his actual practices provided a major impetus for David's adoption of vegetarianism. "If you think you can be an environmentalist and still eat meat, then you are wrong", he insists, reiterating a direct link between ethical consumer practices and identities. The view that the subjects' continuous evolution towards being increasingly ethical in consumption is motivated by a keen longing for the feeling of being true to themselves finds further support in the comments from Joe: "it seemed contradictory to do anything else, I just couldn't justify not being vegan"; Maggi: "there was probably a sort of – I shouldn't be eating fish because I am, in my head, I am vegetarian"; Darren: "I knew it was inconsistent to be eating fish, so I became vegetarian and shortly after that vegan; and Mary: "my self-concept is of someone who is fairly environmentally and ethically aware and I try to be consistent with that … It is more about my own sense of living in a bit more integrity". The urge for authenticity, considered universal to all humans and certainly shared by my respondents, is explained by Vannini (2006, p. 237):

> The basic precept of authenticity is that when individuals feel congruent with their values, goals, emotions, and meanings, they experience a positive emotion (authenticity). In contrast, people experience inauthenticity as

an unpleasant emotion when they perceive incongruence with their values, goals, emotions, and self-meanings.

This insight resonates remarkably closely with the way my respondents feel about their ethical practices, as best exemplified by the next quotes from Lila: "I think I am more coherent with myself when I am more ethical, I am more in harmony with myself"; and Mary: "I feel safe, I feel comfortable in my own skin about it".

The above analysis shows that morally committed consumers approach their ethical compromises from the identity perspective and draw explicit connections between the continuity of their ethical practices and their sense of a coherent, continuous self. While I have provided empirical evidence demonstrating how contradictions and inconsistencies exhibited by ethical consumers create unpleasant feelings of discontinuity and self-contradiction within them, nothing has been said regarding the ways in which consumers deal with these unwelcome, but often inevitable disturbances in their self-concept. Through closely investigating participants' feelings and thoughts about ethical compromises, I have identified a set of ideational strategies by which consumers negotiate the state of intrapersonal moral discord (Frick, 2009) arising from the discrepancies between their declared concerns and their actual practices and mitigate its unsettling effects on their ethical self-image. In this final part of this chapter, I discuss and illustrate each of these strategies with examples from ethical consumers' self-narratives.

Accommodating ethics, negotiating morality: the unethical practices of an ethical self

Identity disavowal

This strategy is best exemplified by Lucy and Joe, the two participants who are strongly committed to a cruelty-free lifestyle and who feel deeply hypocritical at the slightest sign of deviating from their ethical practices. Ethical compromises – both regular and spontaneous, sometimes inadvertent, but often deliberate – form an integral part of their consumption routines: for Lucy, it is a weekly portion of fish and eggs prescribed by a doctor; in the case of Joe, it is an occasional non-vegan snack he treats himself to when unable to resist the temptation. Lucy and Joe mitigate the identity implications of their ethical trade-offs by managing their self-image in a controlled, deliberate fashion: although both self-define as ethical consumers, neither of them lays claim to a vegan identity: "I don't call myself vegan because I eat fish and I eat eggs" (Lucy); "right now I don't tell myself that you're a proper strict vegan at the minute" (Joe). By conceding claims to identities which impose more far-reaching standards of conduct and ethics than those they themselves observe, the respondents avoid disturbing discrepancies between their behaviours and their projected conceptions of self. Interestingly, Joe appeared to be well aware of the ideational process at work here: "not having kind of internalised or externalised identity as a vegan is what stops me from resisting that temptation". His comment suggests that identity disavowal is both an inward and outward-looking strategy – by rejecting particular labels, consumers avoid appearing disingenuous and hypocritical both to themselves and to others.

Vocabularies of motive

In his seminal paper on motives, Mills (1940) develops the notion of "vocabularies of motive" to refer to social discourses from which people draw morally legitimate excuses and justifications for their conduct. It is intended to capture the language that individuals use to explain their motivations and account for their actions in a way that solicits social approval. Vocabularies of motive provide a useful theoretical prism through which to consider consumers' approaches to legitimising the inconsistencies and discontinuities characterising their ethical practices. Here I follow in the footsteps of Grauel (2014) who applies the concept of vocabularies of motive to analyse how mainstream consumers navigate contradictions between representations of responsible consumption circulating in the society and their own consumption behaviours. Grauel identifies a range of motives through which his respondents justified their lack of active participation in ethical consumerism, such as authenticity (in the sense of following one's own preferences and enjoying one's own taste), the needs of the family, and practical constraints. One can see how appeals to these individualistic motives, which have become operative and legitimate in the contemporary neoliberal context, can provide ordinary consumers with meaning-making resources for managing the unethicality of their consumption behaviours.

Akin to Grauel's interviewees, my participants clearly engaged in what Mills describes as "justificatory conversations". Within their accounts, I found many examples of attempts to negotiate deviations from ethical consumption commitments by invoking motives which apparently dominate modern western societies. For example, female participants appealed to the idea of good parenting to justify behaviours that fall outside their ethical standards, as illustrated by Maggi: "my whole focus then was on healthy eating and, you know, eating healthily for my baby … I would have chicken and fish because I wanted to make sure that I was getting the right protein"; Lila: "my daughter is going to enjoy these treats, so I am going to buy it regardless of the packaging and trans-fats"; and Mary: "I was aware of environmental issues of feeding cats food which was fish and meat and stuff that actually was not really environmentally very … it was because my daughter really really wanted her that I gave in." These comments evoke a commonly shared assumption about what it means to be a good parent, namely to lay down your own interests and desires to wholly attend to your children's needs – undoubtedly, one of the "ultimate" and morally legitimate motives in our time.

Another such motive was tapped into by Mary, who invoked the moral value of thrift and reverence for food when explaining why she was willing to eat her kids' non-vegetarian leftovers: "it really seems morally wrong to throw food away". Further, participant accounts were clearly oriented towards excuses in terms of structural barriers to engaging in ethical purchasing and consumption, such as prohibitive prices and unavailability of ethical products. We have already heard the respondents' concerns about the high costs attached to certain forms of ethical consumption, but let me briefly present one more example. In the following quote, Maggi expresses desire to act more in tune with her ethical self: "I should follow my values more and I should buy organic", immediately before invoking financial considerations to excuse her failure to do so: "but then this is a ridiculous price, it is so expensive, so I don't". Very similar vocabularies were deployed by other participants, who described how they "could not afford" (Lucy), "had very little money" (Lila), and "was too poor" (David) to engage in ethical shopping. Framed in these terms, "unethical" choices are least likely to be deplored by segments of the society exhibiting sensitivity towards social inequality and its impact on people's ability to pursue desired courses of action.

In the examples above, consumers displaying ethical contradictions and fragmented selves attempted to establish the moral legitimacy of their actions by interpreting them in terms of involuntary compromises made in the face of other concerns – those which commanded higher moral priority, such as parental responsibilities, which they had no means of escaping simply in virtue of being human, such as health problems and bodily needs, and which they had no choice but to submit to given their conditions and contexts, such as lack of resources to provision ethical goods. As long as unethical acts and activities were framed in these terms, participants felt that the consistency of their moral commitments and continuity of their ethical selves remained unchallenged and undisturbed.

Compensatory reasoning

Participants also displayed the tendency to justify the inevitable deviations from ethical consumption by highlighting their temporary, infrequent nature as well as juxtaposing them with ethical practices they engaged in on a regular basis. Such compensatory logic characterised participants' feelings and thoughts regarding their occasional ethical slip-up, as illustrated by David: "I do so much and I put so much thought into it, and I base my life around these principles for such a long time that I don't feel that bad when I do something wrong"; Joe: "I am working towards a greater good ultimately, it does not matter if I eat the odd sandwich, you know, in the greater scheme of things"; and Lila: "after 15 years of being vegan, I thought it wouldn't harm, it won't do any harm, I'll just try because I feel like it". Another way of boosting one's ethical self-image can be through highlighting ambitions, intentions, and goals which suggest the presence of ethical concerns and commitments even when they are not being acted upon. Maggi, for example, was emphatic that she would like to support a wide range of ethical causes: "ideally I would want fair trade, organic, and local stuff"; and David repeatedly drew my attention to the many ways in which he would be enacting his ethical self if given the opportunity: "if I had an infinite amount of time I would grow more food, I would develop more recipes, I would have time to research all of my foods to find out where everything comes from." This evidence supports the idea that not only actual behaviour, but expressions of preferences and intentions alone can be used by consumers for the purpose of legitimising their self-ascribed ethical status (Grauel, 2014).

Subjective framing

Another strategy, whereby my respondents attempted to avoid clashes of conscience arising from the discrepancy between their assumed ethical consumer identities and their actual practices, is by developing their own subjective criterion of ethical consumption and using it to make sense of and legitimise their contradictory behaviour. For example, Lucy's idea of what constitutes an ethical choice is closely linked to the notions of suffering and harm: "the reason why I eat eggs – I've got friends who keep chickens, and I know that the chickens are perfectly happy, I've seen those chickens, it's not doing them any harm laying those eggs"; "I won't eat farmed fish, I won't eat anything that's been produced in an unnatural way, but if it's deep sea fish, I think – well, at least it had a normal life". Similarly, considerations of harm underpin Solveig's approach to resolving ethical consumption dilemmas that arise before her at the dining table. In the first example, Solveig agrees to taste the cheesecake baked by her elderly grandmother thereby violating her commitment to veganism; in the second scenario, she sets aside her vegetarian principles to eat a seafood dinner cooked by her Nigerian hosts. The notion of harm takes centre stage in Solveig's defence of the moral validity of her conduct:

> I think in a way it is the whole Hippocratic thing, you know, first do not do harm. I know that if I had gone to my grandmother's place and refused the cheesecake that she bought especially because I was coming to visit, I think it would have been more harmful …. And Nigeria – it is a similar thing, but on a larger scale, I think.

Although in both cases, Solveig's actions explicitly violate her commitment to a meat-free diet, they show respect for its underlying moral principle of not doing harm, the ultimate moral benchmark of her consumption and living practices, and therefore do not undermine Solveig's sense of being true to herself. These examples serve to highlight the subjective nature of individuals' understandings and enactments of ethical consumption. While both Lucy and Solveig are guided by considerations of harm in evaluating the morality of their consumption decisions, they part ways when it comes to defining what constitutes harm and whose feelings and wellbeing should be protected from damage: "harm for me would be animal welfare", asserts Lucy; "the priority in this case was really not to hurt people's feelings and not to offend people", explains Solveig. As her comment reveals, Solveig's ethical sensitivity and moral awareness extends beyond animals to include human beings, hence her willingness to deviate from her vegan code for the sake of protecting the feelings of those she sincerely cares about. Moreover, Solveig's reasoning and decision-making about consumption ethics involves priority setting, a process informed by what can be described as a "hierarchy of vulnerability". This subjectively developed idea of who is most liable to suffering and who should therefore be prioritised as the object of ethical considerations is highly consequential for how Solveig construes her moral duties in relation to others:

> If it was my uncle with the cheesecake, I would not have eaten it ... I would have been more happy [sic] to criticise him for not paying attention to what I eat or not eat, because I know that he could take it ... she [grandmother] would be a lot more vulnerable for my criticism than my uncle would have been.

Thus, Solveig's unequal treatment of different family members is based on her subjective evaluation of their dissimilar, in her view, levels of vulnerability and ability "to take it". The hierarchical approach to assuming moral responsibility in consumption is followed by Darren; he, however, uses different criteria for making an ethical choice: "you want to eliminate the suffering of those who are in the most pain, in the worst position". The feeling of moral obligation to take care of those "in the most pain" leads Darren to focus his ethical consumption project on animal life and welfare, hence his strong commitment to veganism.

David and Mary provide more examples of how subjectively constructed ideas and judgments can be exploited by consumers to cope with the threats which contradictory practices pose to their self-image. David, for instance, described how he deliberately adjusted his own beliefs to fit with his meat-eating practices: "I just told myself that by buying local, organic, free-range meat – that is minimising the impact, therefore I can continue as normal as long as I am paying the premium". It is noteworthy that David is fully cognisant that this was a highly intentional and self-serving trick: "I managed to wilfully convince myself that it was not a problem ... because you can twist your morality quite easily like that". Likewise, Mary engaged in deliberate self-deception to feel greater alignment between her proclaimed environmental concerns and her actual consumption behaviour: "I've convinced myself that free-range lamb, for instance, wasn't part of the problem". For a long time, this subjectively constructed belief served to justify Mary's meat-eating habits, the negative environmental impacts of which she remained almost intentionally oblivious to: "I just missed it or did not want to see it". Here we see how by applying their own subjective interpretations, ideas, and understandings of ethical consumption, participants manage to subsume their contravening habits and practices under the term. Designating such practices as "ethical" encourages the resolution of the internal contradiction within individuals: in so far as consumers' conduct does not violate their own ethical standards and their subjective idea of what moral responsibility in consumption entails, it presents no challenges to their ethical self-image and poses no threats to their sense of self-worth. This conclusion confirms and extends the findings of Cherry's (2006) research demonstrating how different definitions of veganism can be used by consumers laying claims to a vegan identity whilst engaging in non-vegan behaviours. How consumers practice veganism, she therefore argues, is directly related to how they define it.

Here, I feel, it is worth clarifying the difference between subjective framing and the vocabularies of motives as ideational tools for managing contradictory and inconsistent instances of ethical consumer behaviour. As examples provided earlier in this chapter demonstrate, appeals to socially legitimate motives and reasons are accompanied by – and in fact stem directly from – consumers' explicit acknowledgement of the "unethicality" of their choices and activities, for which excuses are drawn from the pool of exculpatory motives recognised as morally valid by the wider society. Where consumers attempt to render their self-ascribed ethical status legitimate through subjective ethical framing, they conceptualise and portray their consumption decisions as ethical, only using a subjectively constructed idea of what an ethics-driven, context-sensitive choice involves and which considerations must underpin it. It is also essential to acknowledge that the morality implied by the subjective definitions and frames, which consumers develop and use to protect the validity of their assumed moral identities, cannot be merely arbitrary. Consumers' claims to ethicality, even if they are based on ideas and meanings different from the ones underlying the mainstream view of ethical consumption, require legitimacy, and legitimacy can only be assured by referring to a shared social context (Sayer, 2004). Thus, Solveig would hardly have presented her commitment to avoiding causing harm to others, be it people or animals, as a morally valid ground for claiming an ethical consumer identity if the moral principles of doing no harm and caring about others' feelings were not valued, praised, and commended by society.

Identity disavowal, vocabularies of motives, compensatory reasoning, and subjective framing exemplify some of the ideational strategies (those that are captured in participants' accounts, but there well may be more such techniques) used by ethical consumers to negotiate contradictions and discontinuities permeating their behaviours and practices. In a study of self-defined ethical vegans, Greenebaum also describes a set of accommodating strategies whereby consumers "negotiated the consumption of products that contrasted with their own philosophies, ethics and politics" (2012, p. 132). These accommodating strategies can be seen as part of a broader notion of coping strategies – "constantly changing cognitive and behavioral efforts to manage specific external and/or internal demands" (Lazarus & Folkman, 1984, p. 141) – which has also been used in research on ethical consumers' approaches to dealing with tensions in their behaviours and ethics (Janda & Trocchia, 2001, p. 1216). One important conclusion arising from these studies is that consumers' efforts to justify their ethical contradictions are driven by motives that emerge both from external (social) as well as internal (personal) sources. As Campbell (2006, p. 222) insightfully notes, "individuals have as much need to convince themselves as any observers who may query their conduct". Let me illustrate this with a comment from Joe in which he explains why he persists in observing a vegetarian diet both in public and in the privacy of his home. "I would not want to be that type of person, I would not want to be hypocritical in that way", said Joe when I asked what prevents him from having an occasional "unethical" treat when there is no one around to witness his "failure". "But no one would know, so no one would judge you", I tested him further. "Oh, but I would know, and that would bug me", was Joe's response. This example captures the role of ethical consumption commitments in individuals' self-concept and corroborates the argument made earlier in the chapter, namely that consumers' continuous efforts to negotiate the shortcomings of their ethical practices are driven by a longing for authenticity – the internal state of consistency with their subjective idea of how it is right to be and act for the kind of persons they are. To reiterate, the important conclusion to be drawn here is that consumers' defence of their ethical commitments, projects, and selves, is driven by motives stemming both from internal, personal sources and their relationships with the social world.

On values and practices

One last thing that I want to highlight before closing this chapter. An important finding emerged from the above analysis and discussion which concerns the diversity and divergence of consumers' understandings and enactments of ethical consumption. Its significance extends beyond merely spotlighting the variety of behaviours, practices, acts, and activities which should be acknowledged and accounted for in considering the ethical consumption phenomenon. The inherently subjective nature of people's interpretations and performances of consumption ethics flies in the face of social-practice theorists who persist in the view that "how we understand and actually use these things [consumer goods] will be guided by the organisation of the practice rather than any personal decision about consuming" (Wheeler, 2012, p. 89). The idea that consumption practices are enacted on commonly approved terms fails to account for the variations in the understandings and performances of ethical consumption among participants of both my own research and other cognate studies (e.g. Adams & Raisborough, 2010). While practice-theorists' recognition of consuming agents as "active and creative, constantly reinterpreting social structures and norms within the changing contexts of their lives" (Hards, 2011, p. 25) is a welcome and much needed advancement, to explain the sheer variety of ideas and practices circulating among ethical consumers, referring solely to contextual factors is to acknowledge only one set of causal powers influencing consumer behaviour. As we learned from participants' life stories, people adapt, modify, and refine ethical consumption practices in response to changes both in objective reality and their own subjective concerns. Their narratives tellingly demonstrate that "many lifestyles and types of consumption can be ethically valid, depending on the values, concerns, knowledge, historical background, or social context" (Cherrier, 2007, p. 341), and that "the act of choosing among this wide constellation of possibilities calls for active participation in defining and selecting ethical products, ethical organizations, and, ultimately, ethical consumption patterns" (Cherrier, 2007, p. 322).

Thus, it is the sum total of personal and contextual influences that shapes and moulds ethical consumer behaviours, which may either conform to or defy social expectations and norms. The latter scenario was well exemplified by participants' deliberate departures and deviations from the expectations and norms attached to particular forms of ethical consumption (e.g. environmentalism, vegetarianism, veganism) by mainstream society. Consumers' idiosyncratic enactments of ethical practices are a clear manifestation of their agential ability to relate their moral concerns to the surrounding contexts and identify the most appropriate (that is, subjectively fulfilling and objectively possible) ways of acting upon them. Ethical performances are not homogeneous across consumers precisely because they are informed not only by socially shared ideas, but also by the things individuals truly care about and want to achieve in light of a deep sense of commitment. This statement is echoed remarkably closely in Cherrier's (2007, p. 322) analysis of ethical consumption practices in a postmodern context:

> by becoming active participants in the working of their ethical consumption lifestyles, consumers critically analyze their personal ethical concerns and self-concepts, which initiates customized perceptions and personalized practices of the "good life" and the common good.

This is of course not to deny that personal values are, to a large extent, drawn from shared ideas circulating within society and are greatly influenced by social contexts. Both theoretical and empirical analyses of ethical consumption presented so far support the claim that agential values and practices are socially situated and "are enabled and constrained by the various landscapes in which individuals are embedded" (Hards, 2011, p. 39). This, however, in no way suggests that they, therefore, must be mere reflections of socially dominant ideas and norms which throughout the life of a person remain the sole determinants of how and why she or he engages in certain behaviours. Such a proposition has no chance of withstanding empirical scrutiny, for it homogenises the varied dispositions and meanings that surround ethical consumption, and is unable to account for the multitude of ways in which it is enacted by reflexive, creative, and purposeful agents. Moreover, practice-theorists' assumption that practices produce and co-constitute values (Collins, 2004), and not the other way around, implies that whoever engages in any given social practice is thereby avowing his allegiance to the values that are commonly associated with it. However, people may engage in the same practice, but for different reasons – as Harrison, Newholm, and Shaw (2005, p. 2) point out, "ethical purchases may … have political, religious, spiritual, environmental, social or other motives for choosing one product over another". Furthermore, it is perfectly possible for agents to enact practices without being committed to or even aware of the ideas and meanings attached to them by the wider society – in many cases, ethical choices are imposed on consumers by systems of collective provision and show no sign of emotional investment on the part of the "choosers". To take one example, Jason regularly buys a wide range of organic food products; his preference, however, stems from a belief in the health benefits and enhanced quality of organically grown produce, rather than from a commitment to promoting environmental sustainability. While some people may feel invited to think about environmental values at the sight of Jason's shopping basket and ascribe to him pro-environmental concerns and commitments, his practice of buying sustainable food is driven by self-regarding motives and, therefore, devoid of value content. Similarly, while every purchase of ethical coffee at a fair-trade railway station can be subsumed under the category of ethical shopping and may well have socially beneficial derivatives, a truly meaningful act of ethical consumption requires active awareness of and reflexive identification with its underlying moral cause on the part of an intentional, morally concerned, and actively committed agent. Here, then, lies the difference between merely enacting a practice and expressing a genuine concern.

References

Adams, M., & Raisborough, J. (2008). What can sociology say about FairTrade?: class, reflexivity and ethical consumption. *Sociology*, *42*(6), 1165–1182.

Adams, M., & Raisborough, J. (2010). Making a difference: ethical consumption and the everyday. *The British Journal of Sociology*, *61*(2), 256–274.

Archer, M. (2000). *Being Human: The Problem of Agency*. Cambridge, UK: Cambridge University Press.

Archer, M. (2007). *Making our Way through the World*. Cambridge, UK: Cambridge University Press.

Askegaard, S., & Linnet, J. (2011). Towards an epistemology of consumer culture theory: phenomenology and the context of context. *Marketing Theory*, *11*(4), 381–404.

Barnett, C., Cloke, P., Clarke, N., & Malpass, A. (2005). Consuming ethics: articulating the subjects and spaces of ethical consumption. *Antipode*, *37*(1), 23–45.

Bhaskar, R. (2013). *A Realist Theory of Science*. London, UK: Routledge.

Campbell, C. (2006). Considering others and satisfying the self: the moral and ethical dimension of modern consumption. In Stehr, N., Henning, C. & Weiler, B. (Eds.), *The Moralization of the Markets* (pp. 213–226). New Brunswick, NJ: Transaction Publishers.

Carrigan, M., & de Pelsmacker, P. (2009). Will ethical consumers sustain their values in the global credit crunch? *International Marketing Review*, *26*(6), 674–687.

Cherrier, H. (2007). Ethical consumption practices: co–production of self–expression and social recognition. *Journal of Consumer Behaviour*, *6*(5), 321–335.

Cherry, E. (2006). Veganism as a cultural movement: a relational approach. *Social Movement Studies*, *5*(2), 155–170.

Collins, R. (2004). *Interaction Ritual Chains*. Princeton, NJ: Princeton University Press.

Flatters, P., & Willmott, M. (2009). Understanding the post-recession consumer. *Harvard Business Review*, *87*(7–8), 106–112.

Frankfurt, H. (1988). *The Importance of What We Care About*. Cambridge, UK: Cambridge University Press.

Frick, W. C. (2009). Principals' value-informed decision making, intrapersonal moral discord, and pathways to resolution: the complexities of moral leadership praxis. *Journal of Educational Administration*, *47*(1), 50–74.

Grauel, J. (2014). Being authentic or being responsible? Food consumption, morality and the presentation of self. *Journal of Consumer Culture*, *14*(2), 1–18.

Greenebaum, J. (2012). Veganism, identity and the quest for authenticity. *Food, Culture and Society: An International Journal of Multidisciplinary Research*, *15*(1), 129–144.

Hards, S. (2011). Social practice and the evolution of personal environmental values. *Environmental Values*, *20*(1), 23–42.

Harrison, R., Newholm, T., & Shaw, D. (2005). Introduction. In R. Harrison, T. Newholm, & D. Shaw (Eds.), *The Ethical Consumer* (pp. 1–8). London, UK: Sage Publications.

Jacobsen, E., & Dulsrud, A. (2007). Will consumers save the world? The framing of political consumerism. *Journal of Agricultural and Environmental Ethics, 20(5),* 469–482.

Janda, S., & Trocchia, P. (2001). Vegetarianism: toward a greater understanding. *Psychology and Marketing, 18*(12), 1205–1240.

Laroche, M., Bergeron, J., & Barbaro-Forleo, G. (2001). Targeting consumers who are willing to pay more for environmentally friendly products. *Journal of Consumer Marketing, 18*(6), 503–520.

Lazarus, R. S., & Folkman, S. (1984). *Stress, Appraisal, and Coping.* New York, NY: Springer.

Littler, J. (2011). What's wrong with ethical consumption? In Lewis, T. & Potter, E. (Eds.), *Ethical Consumption: A Critical Introduction* (pp. 27–39). London, UK: Routledge.

Loureiro, M. L., McCluskey, J. J., & Mittelhammer, R. C. (2002). Will consumers pay a premium for eco-labeled apples? *The Journal of Consumer Affairs, 36*(2), 203–219.

McEachern, M. G., Warnably, G., Carrigan, M., & Szmigin, I. (2010). Thinking locally, acting locally? Conscious consumers and farmers' markets. *Journal of Marketing Management, 26*(5/6), 395–412.

Monbiot, G. (2007, July 24). Ethical shopping is just another way of showing how rich you are. *The Guardian.* Retrieved from: http://www.guardian.co.uk/commentisfree/2007/jul/24/comment.businesscomment

Pepper, M., Jackson, T., & Uzzell, D. (2009). An examination of the values that motivate socially conscious and frugal consumer behaviors. *International Journal of Consumer Studies, 33*(2), 126–136.

PriceWaterhouseCoopers. (2008). *Sustainability: Are Consumers Buying It?* London, UK: PriceWaterhouseCoopers.

Sayer, A. (2004). Restoring the moral dimension: acknowledging lay normativity. Retrieved August 30, 2017, from Lancaster University, Department of Sociology website, http://www.lancaster.ac.uk/fass/resources/sociology-online-papers/papers/sayer-restoring-moral-dimension.pdf

Sayer, A. (2011). *Why Things Matter to People.* Cambridge, UK: Cambridge University Press.

Soper, K. (2007). Re-thinking the "Good Life": the citizenship dimension of consumer disaffection with consumerism. *Journal of Consumer Culture, 7*(2), 205–229.

Sudbury, L., & Böltner, S. (2010). Fashion marketing and the ethical movement versus individualist consumption: analyzing the attitude behavior gap. *European Advances in Consumer Research, 9,* 163–168.

Thompson, C., & Coskuner-Balli, G. (2007). Countervailing market responses to corporate co-optation and the ideological recruitment of consumption communities. *Journal of Consumer Research, 34*(2), 135–152.

Vannini, P. (2006). Dead poets' society: teaching, publish-or-perish, and professors' experiences of authenticity. *Symbolic Interaction, 29*(2), 235–257.

Wheeler, K. (2012). *Fair Trade and the Citizen-Consumer.* Basingstoke, UK: Palgrave Macmillan.

Worth, J. (2006, November 2). Buy now, pay later. *New Internationalist.* Retrieved from: http://newint.org/features/2006/11/01/keynote/

8 The inner self in the outer world
The social life of an ethical consumer

Identity is a relation which embraces both our ability to recognize ourselves and the possibility of being recognized by others.

(Melucci, 1996, p. 30)

So far, we have seen how, in an attempt to address concerns which they have designated as the most important in life, morally driven individuals embark on internally conceived, yet externally conditioned projects of ethical consumption, and how they achieve and sustain their desired ethical identities through engaging in alternative modes of provisioning and consuming goods. What one witnesses here is a personal morphogenesis – a process whereby subjects develop into unique persons, each with his or her own idiosyncratic set of subjectively configured concerns which define who they are, that is, their distinct personal identity, as well as what they do – the particular projects and pathways they commit to. In this final chapter, I demonstrate how consumers' inner ethical self, brought about through personal morphogenesis, spills onto their social lives and shapes their relationships with the people around them. Through exploring the links between consumers' ethical concerns and the social roles and positions that they choose to take up or decide to abandon, I provide insight into the dynamics of social identity formation in ethical consumers.

Two specific participants – Lucy and Solveig – take centre stage in this chapter to share their personal experiences of exposing their ethical consumer identities in front of the social world and living them out in public. I bring into focus those aspects of the respondents' lives which reveal the continuous interplay between their commitment to ethical living, on the one hand, and obligations and duties arising from their inevitable positioning as functioning members of society and carriers of certain social roles, on the other. By comparing and contrasting the narratives told by Lucy and Solveig, I demonstrate that consumers' ethical selves may either be allowed to fully define their public persona – how they are towards other people and how other people perceive them, or give way to some other social roles – those more closely aligned with concerns they consider of utmost importance at that particular point in time. Understanding how the ethical consumer comes to occupy different positions in the lives of individuals is a key component of my realist enquiry into ethical consumption for, as Archer points out, "to account for variability as well as regularity in the courses of action taken by those similarly situated means acknowledging our singularity as *persons*, without denying that our sociality is essential for us to be recognisable as *human* persons" (2012, p. 67, emphasis in original). To achieve this understanding, we first need to establish the relationship between personal and social identity from a critical realist perspective.

Archer (2007, p. 17) defines social identity as "the capacity to express what we care about in social roles that are appropriate for doing this". Personal identity, which arises out of human relationships with the natural, practical, and social orders of reality, is inevitably larger and more complex than social identity, which pertains exclusively to the discursive realm. Not only does personal identity encompass social identity – the former is absolutely fundamental to the latter, for it is personal identity that acts as a source of creativity and reflexivity required for shaping individuals' perceptions and enactments of roles in a distinct, inimitable way: "what we need is personal identity in order for any individual to be able to personify a role, rather than simply animating it", argues Archer (2000, p. 288).

Agential performances of social roles are unique precisely because they are produced by the distinct personalities that people develop and bring to bear upon the process of addressing their concerns. At the same time, when personal identity is conceived of as the sum total of an individual's relationships with all three realms of the world, including the social, its formation and existence are dependent on social identity, which determines our social commitments and their relative position within our subjective hierarchy of concerns. Personal and social identities are therefore inextricably linked, though distinctly separate from each other: "both personal and social identities are emergent and distinct, although they contributed to one another's emergence and distinctiveness" (Archer, 2000, p. 288). They are involved in a constant dialectical interchange; within it, three key points of interplay can be discerned whereby a synthesis between one's personal and social identity is achieved to give rise to a unified self.

The first, and in many ways most consequential, moment of interplay is one in which "nascent *personal identity holds sway over nascent social identity*" (Archer, 2000, p. 289, emphasis in original). It is grounded in the crucial role that incipient personal identity plays in determining which roles individuals will consider appealing and worth trying out – their understanding of the range of available possibilities will, of course, always be constrained by circumstance and their inevitably limited experience of the natural, practical, and social realms. However fallible and ill-informed, such exploratory plunges into different roles – both involuntary as well as those that are open for choice – play a key part in enabling subjects to draw "a best-guess sketch of a potential future life" (Archer, 2000, p. 290) and decide which of the roles they have experienced are worth becoming attached to. As Archer stresses, this entire process of exploration of the self and its optimum place in the world by a reflexive subject is possible only because "the nascent personal identity [brings] something to the task of role selection. Otherwise we would be dealing with an entirely passive procedure of role assignment through socialisation" (Archer, 2000, p. 290).

Let me now once again bring the abstract into a relationship with the concrete and apply the above theoretical insights to the lived experiences of two distinct ethical consumers – Lucy and Solveig. In light of what we have just learned about personal and social identities – how they co-emerge, co-evolve, and, ultimately, depend on each other – participants' first attempts at transitioning to cruelty-free living acquire a new meaning, both in the context of their becoming the particular persons they are, as well as their self-expression in public. Lucy's and Solveig's nascent personal identities, defined by emergent concerns over animal rights, propelled them into trying out a particular social role, namely that of a vegetarian, which they deemed appropriate for expressing what they felt they truly cared about. It is their distinct inner self that urged Lucy and Solveig to defy behavioural expectations and norms prevailing in their socio-cultural contexts and engage in social roles they considered appealing and suitable in the light of their moral concerns.

The causal processes and relations underlying the emergence of ethical consumer concerns and their materialisation into specific commitments and practices have been explored at length, theoretically and empirically, in the preceding chapters. Thus, our attention turns to the second moment of interplay, *"where the nascent social identity impacts upon the nascent personal identity"* (Archer, 2000, p. 291, emphasis in original). It is best understood through considering the transformation and change that individuals inevitably undergo as they try out different social roles and begin to amass first-hand experience of their associated material and symbolic benefits and costs. This experiential knowledge informs individuals' decisions about which particular social identity to endorse, which role to invest themselves in, and which social values and goods to appropriate and proclaim as their own. In the process of testing various roles, individuals themselves undergo change, both subjective – by developing a better understanding of their true self: their interests, predispositions, abilities, and talents, as well as objective – by becoming someone with different experiences, resources, features, and skills, which may aid or obstruct the attainment of the next desired position (e.g. transition to a vegan diet is likely to prove less mentally and physically challenging for those who have tried and succeeded in being vegetarian than for lifelong meat-eaters embarking on the very same project).

Participants' experiences of acting as vegetarians in overwhelmingly meat-eating familial and social contexts illustrate precisely this moment of interaction between personal and social identity: inspired and guided by the nascent personal identities, subjects' enactments of the role of a vegetarian provided them with direct knowledge of its concomitant challenges, struggles, and costs. For Lucy, being a sole vegetarian in an omnivorous household resulted in family rifts formed over disagreements about food choices: at some point, her explicit avoidance of meat began to create tensions at the dinner table, with her father becoming increasingly intolerant of her ethical consumption commitments. "He gave me so much grief ... he'd be eating meat and wave it at me, my face, as if to say, come on, don't be silly", recollects Lucy of her unpleasant experiences of family dinners whereby she became aware of a conflict between her chosen role of an ethical consumer and her involuntary position as a daughter expected to share meals with her family. For Solveig too, assuming the role of a vegetarian in the traditional German milieu where meat was considered an essential aspect of cooking came with the risk of undermining her social standing. Although her commitment to a cruelty-free diet was met with respect in her immediate family – both her parents were green voters and held liberal views – it soon started to interfere with the social life she developed and the roles she assumed outside the family context. At secondary school, for example, fulfilling her responsibilities as an ethical consumer came at the expense of her ability to lead a full social life on a par with other teenagers: "after school you go for kebab with your friends, you go for burgers with your friends, and if you don't eat meat – you don't fit in", she recounts. Overwhelmed by the sacrifices involved in identifying as vegetarian, Solveig engaged in a reflexive revision of her prior decision to invest herself in this social role – on second thoughts, she concluded that abstaining from meat "wasn't cool anymore" and "just wasn't the thing you do".

Both for Solveig and Lucy, therefore, initial experimentation with the role of an ethical consumer became a valuable learning experience. Through it, they acquired direct knowledge of the symbolic costs associated with identifying as non-meat eater in their given social contexts: for Solveig, the damage from her commitment to vegetarianism included a reduced quality of social life and a fear of losing her social status; for Lucy, bringing her ethical self to the dinner table resulted in a strained relationship with her father. Equipped with this newly gained knowledge, participants engaged in reflexive deliberations as to whether and to what extent they were prepared to invest themselves in the role of a vegetarian. What both of them had to ask is how important, desirable, and worthwhile it was to assume this particular social identity and make it their long-term, if not full-time, occupation? Similarly placed, their initial role choice being guided by the very same moral concerns, and faced with a similar challenge in the form of compromised social worth, they nevertheless arrived at different outcomes (because, of course, agential decisions are not simply externally dictated, but are a product of subjects' reflexive deliberations through which objective influences are filtered (Archer, 2007)). Lucy decided to continue in the role of a vegetarian, its toll on her family life notwithstanding. Such a choice was possible only due to her being an active and intentional agent, capable of determining what matters to her and what matters to her most of all. Having designated concerns over consumption ethics as her ultimate concerns, Lucy endorsed the role of a vegetarian as most suitable for expressing what she cares about and who she, therefore, is and wishes to be. This was a consequential decision both for her inner and outer self: on the one hand, she reaffirmed her self-concept as an ethical consumer; on the other, she also allowed it to become a key source for her social identity. In the context of family relations, this has been reflected in Lucy's decision to stop sharing meals with her father and eventually move out of the family home.

Solveig took a different path – overwhelmed in her role as the only vegetarian among carnivores, she decided to withdraw from her commitment to a meat-free diet for the sake of her social concerns, namely those over her status among peers and her sense of belonging with them. Such rethinking and refinement of subjective commitments is an essential part of the inner workings of each individual – as we gain enhanced understanding of our inner self and our social roles, we revisit and re-write our life projects to achieve the best possible match between our personal and social identities. This is what internal conversation essentially is – "a dialogue about the kind of person an individual believes they want to be" (Archer, 2000, p. 290), which undergoes constant revision in the light of further knowledge gained through experience and reflexive deliberation. Thus, it is an urge to define who they are and how much of themselves they are prepared to invest in the social realm that prompted Lucy and Solveig to review their initial choice of vegetarianism as their primary occupation and the locus of their self-worth. Since the subjective drivers behind consumers' endeavours to succeed in fulfilling their ethical projects have been explored and discussed in the previous chapter, Lucy's chosen course of action – to carry forward her commitment to ethical eating – should not leave the readers bewildered. At the same time, Solveig's wilful decision to recede from the position of an ethical consumer to accommodate the demands of her other social roles poses questions that have not yet been answered.

This brings us back to the book's central concept, ultimate concerns. What Lucy's and Solveig's accounts of their lives – as distinct persons, ethical consumers, and social subjects – reveal unambiguously is that they are reflexive and evaluative human beings, each with her own unique set of concerns that pertain to the different realms of the world and together determine which commitments and projects they pursue and which ones they avoid. Previously, the idea of concern as the ultimate motivation behind every action provided an effective framework for understanding the subjective forces underlying ethical consumers' commitments and practices, as well as inconsistencies, discontinuities, and contradictions in their behaviours. Building upon these insights, this chapter explores how individuals' concerns affect their choice of social positions and roles. Through inquiring into Solveig's personal story, I intend to demonstrate the role of social concerns in prompting morally committed consumers to retreat from their previously chosen ethical positions and begin the search for a more satisfying and sustainable source for their social identities. This need not be a difficult task, for, in explaining her reasons for abandoning a vegetarian diet, Solveig explicitly refers to her changing concerns and shifting priorities. The following comment highlights this very vividly: "there wasn't a conscious decision – oh, I want to fit in, I want to eat meat again, it was more – I don't care enough anymore".

Although Solveig seems to deny the relationship between her social concerns and her withdrawal from ethical consumption, I argue that it is precisely the desire for social acceptance and the longing to fit in that prompted Solveig to relinquish the role of a vegetarian. Consider the following quote, in which Solveig describes her bitter feelings of social loneliness and exclusion: "I didn't fit in well anyway – I've always been a very bookish child, a quiet child, and I wasn't cool, and

I wasn't pretty ... but, of course, when you're that age, that's a terrible thing ...", which she relates, unprompted, to her decision to return to meat-eating, "because everyone does it, you might as well go along with what everyone does because, perhaps, it's going to help more or less". Solveig's heartfelt comments reveal the underlying cause responsible for unsettling her ethical food practices. A growing tension between her commitment to a meat-free diet, on the one hand, and the intense desire to fit in, on the other, provided the impetus for Solveig to reassess her project of ethical consumption against her social concerns. In light of the knowledge that being a vegetarian comes at the cost of damaged social image and poor sense of self-worth, Solveig decided to renounce this role and let her ethical self recede into the background. This moment marks the beginning of a lifelong struggle between Solveig's ethical and social concerns; depending on which side will be claiming the victory at different points in time, the role played by the ethical consumer in Solveig's social life will vary from forceful to negligible.

In this analysis, nothing seems to go off script which by now we are well-familiar with: individuals continually review their subjective concerns and ensuing commitments in terms of their appeal, worth, and associated costs and adjust their courses of action to reflect the change in priorities, if any takes place. What marks the above scenario out from those we considered before is the nature of concerns that have interfered with the subject's pursuit of ethical eating and made the identity of ethical consumer too constraining and costly to carry in public. Solveig's ethical self was surrendered to the demands of the social positions which she was so eager to occupy and which at that particular stage in her life became the locus of her self-esteem. It was social concerns – the desire to share the norms, behaviours, and practices of her chosen in-group – that prompted Solveig to discard her vegetarian image and establish a new social self. Viewed through the lens of Goffman's (1963) theories of performance and stigma, Solveig's deliberate endeavours to disavow her vegetarian identity can be interpreted as an attempt to avoid social stigma through the management of "problematic" or "spoiled" identity. Goffman's definition of stigma as "the situation of the individual who is disqualified from full social acceptance" (1963, p. 9) and his emphasis on the key role of social perceptions and attitudes in turning a particular feature into a stigmatising factor resonates very closely with Solveig's experience. As Goffman's stigmatised person, Solveig too strove for social recognition and engaged in identity management to fulfil her social needs. Unlike those who experience stigma due to traits that cannot be easily hidden, changed, or removed, such as race, disability, gender, or age, Solveig's non-conformist identity could be easily "remedied" by re-engaging with mainstream eating habits – the route Solveig chose to go down.

Identity management has become an integral part of all Solveig's subsequent life as an ethical consumer. Her pursuit of a cruelty-free lifestyle has been an uneven journey, with her ethical practices being repeatedly sabotaged by the social concerns laying claims to the top position in her hierarchy of priorities. On more than one occasion, Solveig chose to digress from the moral pathway set out by her distinct personal identity, and although concerns over animal rights have always remained on her moral register, expressing them through appropriate social roles proved possible only when there was no cost to her social image in making this choice. For instance, Solveig's commitment to vegetarianism underwent a revival when she got involved with the Gothic movement and the norms, attitudes, and types of behaviour accommodated within it. The following quote highlights these important social influences on Solveig's practices:

> I started to dress differently, started to go to different clubs, hang out with different people, I started university, I started going out with my now husband, and I became a vegetarian again. Mainly because people I spent time with were vegetarians and vegans, and I just thought – yeah, there was something, you know, back then, that is true.

Thus, in a context where alternative consumption behaviour was not merely tolerated, but widely encouraged and practised, Solveig was able to restore her commitment to vegetarianism while causing no harm to her social reputation and sense of self-worth. Vegetarian identity was no longer an obstacle on her way towards social inclusion – to the contrary, it became a means to build up a positive image and achieve recognition in her new social circle. Likewise, social influences have been key in encouraging and facilitating Solveig's transition to veganism. The preceding chapter offered a detailed analysis of the subjective drivers (specific moral concerns and their accompanying emotional commentaries) and some important objective factors (e.g. increased environmental awareness and availability of alternative foods) that motivated her to go vegan. What has not yet been duly acknowledged, and what I am now able to argue in light of the above discussion is that such a commitment proved possible because at that specific point in life it posed no threat to Solveig's social standing and did not undermine her ability to succeed in her other roles. Assuming the role of a vegan was considered a worthwhile undertaking by Solveig not only due to its perceived moral worth, emotional appeal, and practical feasibility, but also because a vegan lifestyle could be easily reconciled with her immediate social context. As Solveig tells us, her first vegan experience occurred whilst she was taking part in the student protests against tuition fees – it was there that she met practising vegans and got introduced to vegan eating. Surrounded by like-minded individuals driven by the same moral concerns, Solveig was able to engage in the desired consumption behaviour with no risk of being rejected, ridiculed, or left out. In a social context where there was no room for concerns over being different and transgressing the norms because being different itself was a norm, she could finally take up the role of an ethical consumer and personify it in her very own unique manner.

However, there is more to be inferred from Solveig's story than merely the enabling and constraining effects of social environments on the ethical consumer inside her. What Solveig has gained through experimenting with vegetarian and vegan identities is the kind of knowledge of the self and the social roles she is drawn to which one needs to be able to define who he or she is and how he or she wants to be seen by others. This is exactly where the third moment of interplay between personal and social identity – that which leads to the final synthesis between them – occurs. On the one hand, we have an objectively and subjectively different person (whose newly gained self-understanding and experiential knowledge of a particular role impact upon her personal identity); on the other hand, it is this person who now has to decide whether this role is a satisfying one and how much of herself she is willing to invest in it. Arriving at this decision requires reflexive effort whereby the subject weighs the demands of the role against those of all other positions he or she occupies in society (for most of us usually hold several social roles at any given time in our lives) as well as against the rest of his or her concerns – those that emerge outside the social order – and ensuing commitments. In this process, Archer (2000, p. 295) argues, "social identity becomes defined, but necessarily as a subset of personal identity". The end result of this synthesis is a personal identity "within which the social identity has been assigned its place in the life of an individual" (Archer, 2000, p. 293).

That social identity represents but one of the many aspects of a person's identity explains why, although both Lucy and Solveig identify as ethical consumers, their social selves take different paths. In what follows, I will demonstrate how by prioritising their ethical and social concerns – each in her own preferred way – Solveig and Lucy achieved final synthesis between their personal and social identities, and what these mergers entailed for their ethical self. Let us first consider Lucy's close, lifelong alliance with her inner ethical consumer who, as we will see shortly, has been allowed to define not only her personal identity, but also her social life and relationships. That Lucy chose to invest her self-worth in the role of an ethical consumer is evident from the way it dominates any other social positions she takes, either involuntarily or through wilful, deliberate choice, as she makes her way through life: a daughter, a partner, a friend. Lucy's own comments suggest that for her, interpersonal relationships almost never gain prominence over commitment to ethical eating: "it never came up as a question, you know, I wouldn't be with a guy who made me eat meat, and if I've got friends who are insisting I eat what they eat, then I don't eat with them". One particular episode in Lucy's life story highlights her priorities especially well. It took place during Christmas dinner hosted by Lucy's mother who, being mindful of her daughter's ethical attitudes, served free-range chicken in the hope that Lucy would share the meal with the rest of the family. Lucy, however, refused to deviate from her ethical food habits, thereby causing emotional distress to her mother and, consequently, to herself: "I felt really guilty because she was upset, you know, I don't want to upset anybody".

The feeling of guilt experienced by Lucy suggests that, first, she realised that by refusing to cooperate with her mother's genuine efforts to ensure a smooth and enjoyable family dinner she essentially failed in her roles as an obedient daughter and polite guest; and, second, that she was not completely indifferent to how well she performed in these roles. What one has to understand to accept this inference is that social affectivity – our emotional responses to social approbation and disapproval – arises from our acknowledgment of the significance and relevance of the situation at hand: "there cannot be any sense of remorse without the personal acceptance that I have done something wrong", as Archer

(2000, pp. 216–217) so rightly points out. Lucy's sense of regret over hurting her mother's feelings indicates that family bonds do hold a prominent place among the things that she cares about, since "for social evaluations to matter – and without mattering they are incapable of generating emotionality – they have to gel with our concerns" (Archer, 2000, p. 219). Yet, this place is apparently not as large as that designated for commitment to ethical consumption. By choosing to subjugate concerns over keeping peace in the family to concerns over ethical eating and by prioritising the role of a vegetarian over the roles of respectful daughter and guest, Lucy has endorsed the right of the ethical consumer to define both her personal and social identities. Her story illustrates very clearly the process whereby the social aspect of the ethical-consumer identity forms and rises to the surface – how people experiment with personifying the role of an ethical consumer and thereby learn about its associated costs and rewards, how they review their tentative role commitments in light of the newly acquired knowledge of the self and the social world, and how they approve this particular role as most appropriate for expressing what they care about and what kind of persons they are.

Another possible scenario is that some other roles become the locus of a person's self-worth, in which case the position of ethical consumer is likely to be explicitly taken only on those occasions when it is "safe" do so from the viewpoint of social concerns. This scenario is well exemplified by Solveig. The ease with which she relinquished the role of a vegetarian whenever it clashed with the demands of other social positions she occupied or was eager to reach suggests that the ethical consumer was not assigned the key role in guiding Solveig's relationships with the social world. Although over the past several years she has become increasingly more persistent in sustaining a cruelty-free lifestyle, her project of ethical consumption is riddled with compromises made for the sake of appeasing her social concerns. One of the most illuminating examples of Solveig's preparedness to give up the position of ethical consumer to succeed in her other social roles relates to her passion for softball. An avid softball player and fan, Solveig knows the value of having a quality leather glove and has always been eager to get one of her own. For obvious reasons, fulfilling this desire required a surrender of her vegan principles, and Solveig proved willing to do so. This was a conscious, active decision which she justifies in the following way:

> It is sheer impossible to get faux leather gloves – they are ridiculously expensive and I just would not be able to afford it. And not having your own glove is not really good, you can borrow them from, I don't know, the sports kit, for example, but the glove shapes around your hand and you just play better with your own glove. So yes, it's a compromise, but one I am willing to take.

In this scenario, Solveig's concern over her worth as a softball player came into conflict with her commitment to veganism and its concomitant ban on the consumption of leather goods. She could, as she tells us, buy a synthetic glove or borrow one from a sports kit – the first option, however, was prohibitively expensive, while the second meant compromising her playing abilities. Both ethically satisfying solutions, therefore, came with a cost – material in the first case, symbolic in the second – which Solveig was unwilling to bear. Unable to reconcile her competing concerns, she decided to sacrifice her ethical principles for the sake of achieving success on the softball field.

Solveig's preparedness to recede from the position of ethical consumer to meet social expectations and norms was equally clearly demonstrated by the two episodes discussed in the previous chapter – those where she consciously compromised on being a vegetarian by eating her grandmother's cheesecake and sharing a seafood dinner with her Nigerian hosts. We heard Solveig legitimise these ethical contradictions by framing them as attempts to stay loyal to the ultimate moral principle governing her consumption and living practices, namely the principle of doing no harm: "the priority in this case was really not to hurt people's feelings and not to offend people", she explained. Here I want to yield insight into the inner workings behind Solveig's decision to deviate from consumption ethics and consider it in the context of her other social roles. The dilemmas that arose at the dining table required Solveig to arbitrate between the standards of behaviour she was supposed to uphold as a practising vegetarian and the norms, demands, and pressures attached to the roles of granddaughter and guest. The route Solveig took in both cases reflects more than her commitment to minimising the amount of suffering in the world, for there were not only the feelings of others at stake – there were also her reputation, social image, and public esteem, as the following comments reveal very well: "I wouldn't have felt comfortable sitting there saying – "oh, no, I am not going to eat that cheesecake"; "I am a guest there ... not accepting it [the food] would have been rude beyond belief". It is this predictive analysis of the emotional (the feelings of discomfort, embarrassment, guilt) and symbolic (diminished social worth as a result of appearing rude, disrespectful, and lacking in manners) costs of transgressing social expectations about the proper way to behave in the role of a guest that informed Solveig's decision to digress from her intended ethical path.

Solveig's respect for social norms is further reflected in her disapproval of a radical approach to addressing ethical problems, such as illegal activities of animal liberators. While she wholeheartedly supports the causes pursued by environmental and animal rights activists, she condemns their militant tactics for being socially subversive:

> I know that they are probably doing the right thing, but I have a feeling that there are other ways to get to these goals, more socially acceptable and effective ways, and they always seemed to be a bit extreme to me.

Although at present concerns over social acceptance no longer interfere with Solveig's ability to act in ways that express her ethical consumer identity (an important enablement stemming from a socially diverse environment in which she is placed), they continue to exert influence on her consumption behaviour by determining the range of activities she chooses to engage in or avoid. For instance, farmers markets – a trendy destination for shopping and socialising among ethically minded consumers – are disfavoured by Solveig because of their particular ambience and clientele:

> It is mainly young professionals and young families, and it is very hipster and very cool to buy organic food, and I just – it just annoys me so much. I don't feel … I don't like the atmosphere because I feel it is a bit pretentious … I really wouldn't feel comfortable among these people.

Solveig's unpleasant affective response to farmers markets, where she feels annoyed, uncomfortable, and out of place, stems from the desire not to lose the feeling of belonging and being comfortable in her own skin: "I don't feel like I am the kind of person that fits in there", she says. Evidently, social concerns and their accompanying emotional commentary constitute one important factor that prevents Solveig from embracing farmers markets as a source of ethical food.

My analysis reveals that for Solveig, social pressures, evaluations, and norms consistently prevail over the requirements set by the role of an ethical consumer, suggesting that the latter has been denied the right to define Solveig's social identity and that instead her other social projects – family, partnerships, friendships – have become the locus of her self-worth. Because Solveig's role enactments are a means of addressing her social needs, which apparently occupy high positions in her subjective hierarchy of concerns, she cannot stay indifferent to how well her performances are received by others. As Archer reminds us, "it is because we have invested ourselves in these social projects that we are susceptible of emotionality in relation to society's normative evaluation of our performance in these roles" (2000, p. 219). Similarly, it is because for Solveig being a social subject, a functioning member of society, and a valuable part of her chosen in-groups comes before being an ethical consumer, that contradictions and compromises pervading her consumption behaviour do not produce the same devastating effects on her self-esteem as does failure to succeed in those other social roles. Unlike Lucy, whose self-image and self-respect are inextricably bound up with the continuity and consistency of her ethical consumption commitments, Solveig does not feel dishonest or hypocritical whenever she fails to comply with the requirements of a cruelty-free lifestyle. In part, she prevents herself from feeling inauthentic by framing her ethical compromises as a tribute to the higher moral principle of not doing harm – an accommodation strategy discussed in the previous chapter. But it is also because social concerns rank very highly among things Solveig cares about and because her self-concept depends more on her success as a social being than as an ethical consumer that inconsistencies in her consumption behaviour are tolerated much more readily than they would have been otherwise. That being a morally conscious consumer represents but one aspect of Solveig's self-image is revealed very clearly by the following quote: "it is part of who I am, yes, but it is not … Like when I meet someone I wouldn't say – hi, I am Solveig and I am vegan … It's not the first thing I would tell someone".

But if social concerns feature so prominently among those that define Solveig's identity, if social relations and bonds constitute the locus of her self-worth, and if other social roles have become a key source for her social identity, the question arises as to what place the ethical consumer holds within her overall self? For if we fully embrace the idea that "which precise balance we strike between our concerns, and what precisely figures amongst an individual's concerns is what gives us our strict identity as particular persons" (Archer, 2000, p. 221), are we then to assume that whenever Solveig refused to prioritise her commitment to veganism over social expectations, evaluations, and attitudes she thereby wilfully surrendered her ethical consumer identity? Does the fact that consumption ethics shift up and down her subjective hierarchy of concerns depending on how well her social concerns are catered for at any given point in time mean that Solveig is therefore less of an ethical consumer than Lucy, for whom ethical eating and living practices almost invariably take the highest priority? Does the fact that for Solveig being a good granddaughter or a respectful guest comes before being vegan or vegetarian necessarily mean that it comes *instead*? My answer to all of these questions is no, for such a view presupposes a pitifully narrow and impoverished concept of identity for which there is no room in a realist theory. To argue that how we prioritise our concerns is what defines us as persons is not to suggest that whenever subjective or structural forces push us closer towards certain concerns and away from the others, we instantly lose our former identity and put on a new one instead. First, Lawler reminds us that "no one has only one identity; and indeed, those identities may be in tension" (2008, p. 3). The second key point to bear in mind is that people do not simply discard their cherished concerns in response to changing life circumstances, but they continuously reshuffle them in ways that reflect their shifting priorities. This means that identities defined by these deep-seated concerns cannot be simply lost either. These identities, with all their concomitant attitudes and beliefs, may be forced into the shadows by some other concerns that arise before people as they go through life and that may come to be viewed as an utmost priority. Yet, even when certain aspects of our identities are suppressed, they do not cease to exist or exert influence on individuals' intentions and actions. While social relations and bonds may well be a source of Solveig's social identity and the locus of her self-worth, concerns over ethical consumption continue to define what kind of person she is and what she chooses to do. This is because, as already stated, subjects' social identity makes up only one aspect of their personal identity which "intertwines with their sociality, but exists *sui generis* and cannot be reduced to it" (Archer, 2000, p. 196). As Archer (2000, p. 295) explains:

> the social positions we occupy do contribute to the person we become, which is why this is presented as a dialectical process: but the final synthesis is one which finally defines the person as someone with concerns in the natural and practical orders, as well as the social order.

For this reason, Solveig's social concerns neither exhaust all there is to her personal identity, nor govern all her relationships with external reality. Undoubtedly, her social self to a large degree controls when, where, and in what form her inner ethical consumer will reveal itself to the world – indeed, we have seen that for Solveig, the demands of self-presentation often prevail over the urgency of self-expression. Yet, this is not a one-way interaction, and commitment to ethical living plays a key role in determining who she is and how she is towards others. The following comment accentuates these important mutual influences:

> I am trying to be accepting, and I am trying to be caring and loving, and I am trying to live a life that has – that does not have a negative impact on others … for me eating animal products wouldn't really fit in there, it is just a logical consequence.

Following Lawler (2008, p. 39), the perpetual tension between Solveig – ethical consumer and Solveig – social subject may be viewed as a conflict between "the self as an autonomous entity" and "the self as the embodiment of relationships", where the former is guided by concerns over the morality of consumption, while the latter imposes constraints on the extent to which these concerns can be realised through appropriate roles.

My analysis of the trajectories of Solveig's and Lucy's social selves yields a number of important insights that complement the account of ethical consumer identity presented in the first part of the book. First, it illuminates a dialectical relationship between personal and social identity and the ensuing possibilities for the formation of a comprehensive ethical consumer identity, which may or may not be realised depending on the objective contexts and subjective priorities of the concerned individual. This is important because by understanding how one finds a set of social roles in which he or she decides to invest his or her self-worth and how one then arbitrates between the demands of his or her chosen positions, we can understand how the ethical consumer persona gets assigned its particular place in the life of an individual and why these places differ in prominence from person to person. The idea of identity, as defined by concerns that pertain to three different orders of reality, goes a long way in explaining why even the most deeply committed ethical consumers cannot evade or remain completely indifferent to concerns arising in the discursive realm. Let me illustrate this argument with several brief examples from a persistent practitioner of ethical consumption, Lucy. A proponent of cruelty-free eating, she refuses to serve non-vegan food to her omnivorous guests, be it a special occasion or a casual dinner party for friends. Yet, the desire to do well in her role as a host propels Lucy into making a small compromise, so she uses regular milk when preparing coffee or tea for her guests. "It is just being hospitable, you know, I don't want people go away thinking, bloody hell, she put soya milk in my drink", she says, revealing that not only does she recognise the social expectations and norms about how to be a good host, but that she has internalised these beliefs, thereby, becoming susceptible to how others perceive and assess her performance in this particular role.

Lucy's social concerns are further reflected in the way she goes about her ethical principles when playing the role of a guest herself. While she is unwilling to compromise on her consumption commitments at social gatherings and events, she never expects or demands to be specially catered for: "I just don't really want to draw a lot of attention to myself, it's tiresome, and then you got somebody thinking – god, she is a pain in the neck!", she explains.

The tendency to keep alternative food habits as low-profile as possible has been observed by Janda and Trocchia (2001) in a study of practising vegetarians. The authors interpret it as a coping strategy, whereby non-mainstream consumers resolve tensions between individual freedom and social belonging and avoid social judgement and criticism. The following comments from Lucy reveal unmistakably that she too is liable to concerns over social image: "I don't want to irritate people"; "I don't want to be a complete bore". Although these concerns do not take over Lucy's personal or social identity (we have already seen that commitment to ethical living plays a dominant part in determining what kind of person she is and how she relates to the world), they nonetheless exert an influence on the way she behaves towards others. While taking nothing away from her ethical consumer identity, Lucy's deliberate efforts to prevent her social and ethical status from clashing (e.g. by keeping the public attention diverted away from her eating commitments while nonetheless staying fiercely loyal to them) indicate that she too, inevitably, has concerns in the social realm that are part of her human condition and that affect, even if only in minor ways, how she expresses her ethical self.

This concludes my account of the subtle, but highly consequential interplay between personal and social identity, and its relation to the process and outcome of one's formation as an ethical consumer. In this final chapter, my aim has been to explore how individuals arbitrate between their ethical and social concerns thus achieving a distinct identity within which both their ethical and social selves have been assigned their particular places. I chose to focus on Solveig and Lucy because of their contrasting experiences, providing an opportunity to catch the best view of the continuum along which individuals shift their ethical consumer identity in relation to their public persona. My analysis has revealed that despite different outcomes, the process whereby Lucy and Solveig achieved final synthesis between their personal and social identities is driven by the same underlying mechanism, which can be summarised in the following way:

- Under the impetus of their nascent personal identity defined by concerns over consumption ethics, individuals begin to experiment with the role of an ethical consumer. This means not merely performing the role according to a socially supplied script, but personifying it in their own unique manner;
- Through these exploratory role enactments, subjects acquire first-hand knowledge about the costs and benefits of occupying the position of an ethical consumer within mainstream society;
- In light of this newly gained knowledge, individuals engage in reflexive deliberations as to whether the ethical consumption arena might become a satisfying source for their social identity and the locus of their self-worth;
- Depending on how much of themselves individuals decide to invest in the role of an ethical consumer, the latter gets assigned its part in defining their

social lives and relationships – it may become an absolute leader, a mere assistant, or anything between these two extremes;

- No matter which position ethical consumer ends up occupying in people's social lives, it remains an integral part of their distinct identities and continues to exert influence on their intentions and actions for as long as consumption ethics can be found among their ultimate concerns.

This outlined mechanism provides explanatory purchase on both the formation of ethical consumers' social identity and the varying degrees of its visibility – from near-absence to sporadic appearances to full-time presence – in the lives of individuals. Archer's theory of the emergence of social identity, brought to life by an empirical analysis of personal experiences of two distinct ethical consumers, enabled me to identify and illustrate this mechanism at work in ethical consumption. In tracing the formation of social identity in ethical consumers – walking through its key stages and analysing how they connect up to each other – I sought to demonstrate that the inextricable links between ethical consumers' concerns and identities extend over their social selves, and that by committing to more ethical lifestyles people define not only who they are, but also *how* they are towards others. This is a necessary relationship, for in the absence of concerns about the wider environmental, social, and moral impacts of consumption the ethical consumer persona (by which I mean a deliberate and genuine manifestation of a person's true self as opposed to a scripted performance) can neither appear nor thrive, for it is our ultimate concerns that inform what social roles we take up and how we personify them:

> What we seek to do is reflexively defined by reference to the concerns that we wish to realise. Ultimately, that realisation means becoming who we want to be within the social order by personifying selected social roles in a manner expressive of our personal concerns.
>
> (Archer, 2007, p. 88)

Furthermore, the emotional import of concerns over consumption ethics needs to be sufficiently strong for them to prevail over individuals' other inescapable concerns and to get assigned a dominant role in shaping their social identity. It is worth noting, however, that the successful enactment of the role of a conscientious consumer does not automatically mean that active prioritisation of ethical concerns over those pertaining to the discursive order has necessarily taken place. As we learned from Solveig's story, insofar as a commitment to alternative modes of consuming does not threaten the agents' ability to pay heed to their social concerns, the need to actively choose between the two will most likely never arise, and the ethical consumer persona can remain on a par with other social roles assumed by the subjects.

Another important point worth reiterating as the chapter draws to a close is that regardless of their relative hierarchical positions, both ethical and social concerns retain their roles in determining an individual's personal make-up. Relegated to the more private aspects of her life and identity, the ethical consumer remains an important constituent of Solveig's distinct personality, just as Lucy's social concerns continue to exist alongside, although never rising above, her ethical self. To succeed in rendering meaningful ethical-consumer behaviour with all the inherent contradictions, disruptions, and discontinuities, we need to acknowledge that ethical consumers are human beings who have not only moral but social concerns, and who continuously negotiate their relationships with the discursive order to achieve a satisfying balance between their inherent normativity and their inescapable sociality. Because identities are *"embedded within* and *produced by* the social world" (Lawler, 2008, p. 144, emphasis in original), people's idiosyncratic enactments of roles, enabled by their concrete singularity as persons, are improvisations within a scene of constraint, to paraphrase Butler (2004). Likewise, personifications of the role of a conscientious consumer by active, reflexive, creative agents, however unique, are always and necessarily restricted by their immediate social contexts.

This point provides a suitable ending to the discussion we witnessed unfold on these last several pages, for it reiterates the book's overarching emphasis on recognising and analysing both the social (the objective) *and* the personal (the subjective in explaining consumer conduct and human behaviour in general. The importance of this is strongly argued by Archer (2003, p. 25):

> Unless we acknowledge this, we will go far astray by making assumptions that the same constraints and enablements have a standardised impact upon all agents who are similarly placed. Instead, in every social situation, objective factors, such as vested interests and opportunity costs for different courses of action, are filtered through agents' subjective and reflexive determinations. Actions are not mechanically determined, nor are they the subject of a uniform cost-benefit analysis that works in terms of a single currency of "utiles". Rather, it is the agent who brings her own "weights and measures" to bear, which are defined by the nature of her "ultimate concerns".

Acknowledging both the objectivity of social contexts and the subjectivity of consuming agents and exploring the ongoing dialectic between them is indeed a necessary condition for producing a successful critical realist account of ethical consumption and consumer behaviour more broadly.

References

Archer, M. (2000). *Being Human: The Problem of Agency*. Cambridge, UK: Cambridge University Press.

Archer, M. (2003). The private life of the social agent: what difference does it make? In J. Cruickshank (Ed.), *Critical Realism: The Difference It Makes* (pp. 17–29). London, UK: Routledge.

Archer, M. (2007). *Making Our Way Through the World*. Cambridge, UK: Cambridge University Press.

Archer, M. (2012). *The Reflexive Imperative in Late Modernity*. Cambridge, UK: Cambridge University Press.

Butler, J. (2004). *Undoing Gender*. New York, NY: Routledge.

Goffman, E. (1963). *Stigma*. Englewood Cliffs, NJ: Prentice-Hall.

Janda, S., & Trocchia, P. (2001). Vegetarianism: toward a greater understanding. *Psychology and Marketing*, *18*(12), 1205–1240.

Lawler, S. (2008). *Identity*. Cambridge, UK: Polity Press.

Melucci, A. (1996). *The Playing Self: Person and Meaning in the Planetary Society*. Cambridge, UK: Cambridge University Press.

Conclusion

This book has told the story of an ethical consumer. This story began with an individual – an inherently emotional, normative, evaluative human being with an innate capacity for reflexivity, self-awareness, and intentional action. We followed her transformation into a distinct person whose subjective relationship to the world is one of concern over consumption-related ethical problems and her personal role in shaping their outcomes. We then witnessed this relationship translate into a project of ethical consumption which is not merely a set of morally inspired practices, choices, and activities, but also, and most importantly, a tangible manifestation of our subject's most precious commitments and a master narrative about what kind of person she is and wishes to be.

By drawing on the narratives woven together by my participants, I demonstrated that the process of consumer moral conversion begins with the properties of objective reality – circumstances, conditions, and states which exist independently of their identification and evaluation by agents, but which have the potential to become objects of their ultimate concerns. For this to occur, these objectively real properties need to come into play with the subjective properties of individuals – their capacity for moral evaluation, emotional elaboration, and reflexive deliberation. Opportunities for cognitive and affective engagement with matters of potential concern are created through glimpsed experiences – direct or mediated encounters with the world of production, which allow people to comprehend as well as feel the ethical consequences of individual and collective consumption behaviour. Before materialising in concrete actions and practices, consumers' incipient ethical concerns will have to be sifted through and assessed in terms of their emotional appeal, moral worth, and potential to become a lifelong commitment. This reflexive effort takes place during the internal conversation – an incessant cycle of inner self-dialogues via which ethically driven consumers work out their relationship to the world and their moral duties towards it. Through a unification of feeling and thought, the internal conversation leads aspiring ethical consumers to their ultimate concerns, commitments, and suitable ways to fulfil them. It is this complex and intricate inner process at the end of which consumption ethics become a person's ultimate concern, and environmentally, socially, and morally responsible living – his or her most precious project that produces an ethical self out of the raw materials of consumers' mental and emotional lives.

Through telling the story of consumer moral conversion, this book refines and extends what we currently know about ethical consumption as a social phenomenon and ethical consumers as individual agents in two major ways. First, the book presents a theoretically driven, empirically grounded account of the process and mechanism underlying the development of the ethical consumer identity, a glaring missing element in the extant literature on identity relations and processes surrounding ethical consumption. Grounded firmly in the view of humans as "beings whose relation to the world is one of concern" (Sayer, 2011, p. 2), this account goes beyond merely highlighting the presence of specific moral concerns in ethical consumers' attitudes and behaviours. Not only does it empirically demonstrate that ethical concerns, commitments, and practices are central to the making of a conscientious consumer, but it also provides the theoretical explanation for their causal inter-relationship and its role in the production of consumers' ethical self. Further, it identifies reflexivity as a central generative force behind ethical consumer

a   concerns, for it enables individuals to engage – both cognitively and emotionally – with the social, environmental, and moral issues around consumption and designate them as their ultimate concerns;

b   commitments, as it is only by reflexively considering their concerns in relation to their contexts and vice versa that consumers can devise projects that are both subjectively satisfying and objectively feasible;

c   practices, since sustaining a consistent, coherent, and stable ethical lifestyle requires continuous reflexive monitoring, evaluation, and negotiation of structural enablements and constraints by an active, creative, intentional agent; and

d   identities, for it is the concerns we designate as our ultimate and the commitments we dedicate our lives to that define who we wish to be and who we become.

This reveals the pivotal role of reflexivity in enabling individuals to achieve and sustain desired ethical consumer identities within the constraints imposed by their subjective needs and objective contexts, alongside highlighting its function as a key mediating mechanism linking consumer's inner self with the outer world. The concept of reflexivity has, therefore, been critical in allowing me to integrate agential subjectivity and structural objectivity into a single story and provide a theory-driven account of their interplay in the context of ethical consumption.

It is here that this book makes its second key contribution to the field of consumer studies. By exploring how structural properties and their ensuing influences on agential intentions and practices interact with the capacity of individuals to define their own life and identity pathways, it begins to redress the biases inherent in the current approaches to conceptualising and studying consumer behaviour. Theoretical and methodological tools necessary for developing a dialectical perspective on consumption have been provided by critical realism, which puts a distinct emphasis on researchers' ability "to distinguish sharply, then between the genesis of human actions lying in the reasons, intentions and plans of human beings, on the one hand; and the structures governing the reproduction and trans-formation of social activities, on the other" (Bhaskar, 2010, pp. 75–76).

This analytical dualism has been vital in enabling me to explore and explain how consumers' voluntary commitment to ethical living intertwines with their involuntary placement in the three different realms of the world. It has been revealed very clearly that every project of ethical consumption, while stemming directly from the ability of individuals to deliberate upon the world, define what matters to them in it, and devise appropriate courses of action, is always con-tained within the enabling and constraining properties of the objective contexts in which it unfolds. Ultimately, my analysis demonstrates "*how* structure actually does impinge upon agency (who and where) and *how* agents in turn react back to reproduce or transform structure" (Archer, 1998, p. 371, emphasis in original) in the context of ethical consumption.

By teasing out the dynamics of identity formation in ethical consumers, chart-ing the key steps in this multi-dimensional process, and revealing the main driv-ing forces behind it, this book has shed new light on what becoming and being an ethical consumer entails and what it means to the subjects involved. This brings us a step closer to a more complete understanding of ethical consumers as indi-vidual agents and as social actors. This is not to suggest that the ethical consumer identity can be construed in terms of a fixed set of predefined features and pre-determined actions that anyone who identifies in this way must be able to match. That the assumption of an essential and unchanging identity is deeply mistaken becomes immediately apparent as soon as one begins to consider just how varied individuals' conceptualisations and enactments of ethical consumption are. No two ethical consumers I met through my research could possibly be squeezed into one personality type, unsurprisingly so given that they all come from dif-ferent backgrounds, are guided by their own subjectively chosen and uniquely configured concerns, and face diverse and distinct enablements and constraints. Yet, despite their undeniable singularity, my participants share some common characteristics representing their essential properties as human beings and, more specifically, as ethically concerned and committed consumers.

First, the ethical consumer of my research is an emotional, moral, and value-driven human being. Her figure confronts the rationalist portrayal of ethical con-sumption as a self-serving pursuit of a "risk discounting and profit-maximising bargain-hunter" (Archer, 2000, p. 55), "who knows the price of everything and the value of nothing" (Archer, 2000, p. 4). While her project of ethical consump-tion is not void of rational considerations – for all agential commitments have to be evaluated not only in terms of their moral worth and emotional appeal, but also in terms of their practical feasibility and potential to become a lifelong undertak-ing – it is guided not by a cold-blooded calculation of the ensuing losses and gains, but by emotion-imbued internal self-dialogues about her deepest concerns. She does not conform to the model of rational, inward-looking, preference-driven individuals who do good not in order to be good, but rather to feel good about themselves. Instead, she fits perfectly the image of a human being whose projects are inspired and guided by concerns that are "not a means to anything beyond them, but are commitments which are constitutive of who we are, and an expression of our identities" (Archer, 2000, p. 4).

The ethical consumer of my research is a reflexive, evaluative, and self-aware individual. She is not a Bourdieusian subject whose personal dispositions are merely a reflection of objective positions and whose ways are guided by habitus which is nothing more than internalised social structures and facts. For her, the absence of self-awareness is simply not a viable option, because it is through constant reflexive assessment of the self and its relationship to the world that she arrives at her ultimate concerns and her precious commitments. She cannot concede her reflexive capacities, for they are essential for her ability to continuously monitor and evaluate her subjective concerns and objective contexts and adjust her ethical project in response to anticipated and actual changes. The ethical consumer of my research is, therefore, "the author of his own projects in society" (Archer, 2003, p. 34). He is not an over-socialised subject whose qualities are supplied by the social world, whose values are not reflexively developed, but are mere derivatives of what is considered important or worthy in the wider society, and whose behaviour and practices are determined by a set of commonly accepted values and norms. His idiosyncratic sense-making and performance of ethical consumption overturn the assumption of the primacy of social discourses over personal thoughts; they attest to his being the ultimate cause of his actions and practices, shaped as much by his own creative contributions as by the effects of socio-cultural contexts.

The ethical consumer of my research is an active, intentional agent. Her example defies the idea of a "constructed" consumer whose practices are neither wilfully chosen nor actively shaped by herself, but are dictated and scripted by a set of strategically oriented actors in pursuit of their own vested interests. Far from falling prey to the neoliberal project of the governing of consumption (Barnett, Cloke, Clarke & Malpass, 2010), she exercises her capacity for autonomy in reflecting and making active choices about what, where, and how to consume. Inevitably, however, the ethical consumer of my research is also "a wavering, suspicious, ambivalent 'consumer' juggling their choices amidst competing knowledge claims" (Adams & Raisborough, 2010, p. 270). Her project of ethical consumption is riddled with precariously reconciled contradictions, discrepancies, and discontinuities, for in the absence of definitive guidelines about how to improve human, animal, and planetary wellbeing ethical consumption discourse remains fluid, meanings – open for interpretation, practices – subject to change.

Finally, the ethical consumer of my research is a contextually embedded individual whose ability to pursue his preferred course of action is contingent on the nature of his relationships with the three different realms of the world. His inner self *is* "a *reflexive* project" (Giddens, 1991, p. 32, emphasis in original), but one that is always externally conditioned under objective circumstances, and while he is free to choose his own identity, he has limited power to choose the conditions under which he will make it. His agential subjectivity is always juxtaposed against but embedded within structural objectivity, and it is in their continuous interplay that the shaping of his consumption practices and the forging of his ethical identity takes places. The ethical consumer of my research is therefore like "a sculptor at work fashioning a product out of existing materials using the tools available" (Archer, 1998, p. 360). He can be likened to an elephant rider, to borrow Haidt's (2012) metaphor, who wants to follow a particular route and is trying to steer the elephant in the desired direction, but has only limited control over the mighty animal. Like little riders on the back of a giant elephant, ethical consumers are striving to pursue their chosen consumption and living practices, but always find themselves constrained by external powers.

Both these main drivers of ethical consumer behaviour have exposed themselves to my view – the researcher's "gaze" that I brought to the field. When observing participants while they were shopping for groceries, I was looking directly at the riders – active, deliberative, intentional agents, driven by a deep sense of concern and responsibility over consumption ethics, seemingly mindful of every small choice they made, rarely failing at providing a reason for why this or that particular product earned – or was rejected – a place in their shopping trolleys and baskets. Then, having learned their life stories through formal and casual interviews, I met face to face with the elephant. Comprised of participants' family backgrounds, educational inputs, cultural exposures, social pressures, and structural opportunities, it symbolises a complex interweaving of diverse social factors that have been shaping and moulding participants' pathways leading them to become the particular persons they are. This powerful elephant is not just a product of past experiences – it is caught up in an intricate web of objective conditions which exert a strong influence on individuals' ability to act upon their ultimate concerns, to fulfil their largest and most cherished commitments, and to achieve their desired identities. This eloquent metaphor allows us to recognise in ethical consumption an arena of a continuous interplay between the riders – active, creative, purposeful agents in pursuit of reflexively produced courses of action, and the elephant – a dynamic set of external influences, shaping the general direction and every little turn of individuals' lives.

This depiction is not an attempt to combine the behaviours and traits of ethical consumers into a rigid personality profile, nor is it intended to endorse the view of an ethical consumer as having a fixed essence that is his identity. To suggest that there are universal, intrinsic features that make us beings "for whom things matter" (Sayer, 2011, p. 99) and who conduct their life accordingly, is not to deny the prodigious diversity of people's identities and the ways in which those identities can be expressed and communicated to others. As Sayer rightfully notes, "making claims about the particular capacities of human beings does not mean that they are all manifested equally or in the same way everywhere and never change" (2011, p. 104). The assumptions that identities have a pre-determined or fixed essence and that individuals' performances are passively scripted with social expectations and norms have been repeatedly refuted in this book, the idea of reflexivity being the key in explaining why none of them can plausibly be regarded as holding true for active, intentional, and self-aware agents that all of us are. Yet, my analysis brings to light some fundamental characteristics of ethical consumers – certain powers, properties, and propensities which they have in common and which explain what it is about people that makes them moral agents capable of other-regarding concern, unselfish commitments, and self-sacrificing behaviour. These admittedly universalist assumptions about human nature are not only permissible, but necessary, for without them we cannot hope to understand human behaviour: "it is hard to say anything much about people or indeed interact with them without presupposing something about what they have in common" (Sayer, 2011, p. 106). Such universalism, Sayer explains, need not imply uniformity – moreover, our differentiation is enabled *precisely* by our essential commonalities. Indeed, it is what ethical consumers have in common – the propensity to be concerned and care about others, capacity for emotionality and cognition, ability to devise intentional courses of action, reflexivity, creativity, self-awareness – that is responsible for the idiosyncrasies, divergences, and unique manifestations of their identities.

Finally, in interpreting engagement in ethical consumption in terms of the pursuit of desired identities, I do not mean to portray ethical consumers as self-centred individuals chasing their own personal goals, or strip ethical consumer behaviour of its underlying altruistic and other-regarding motives. As Sayer (2011) insightfully points out, people do what they do not simply because they want to do it, but because certain things really matter to them, because as inherently moral beings they genuinely care about the lives and situations of others, and because their own happiness and wellbeing directly depend on whether the objects of their concern prosper or languish. That we care about some particular things and not others is precisely what defines what kind of persons we are: consumers' commitments to ethical living are a direct expression of their other-regarding concerns which, in turn, are extensions and reflections of their compassionate, caring, benevolent self. Through engagement in ethical consumption, individuals achieve, maintain, and express their authentic identities; yet, their ethical choices and activities are not mere instruments for the construction and presentation of self, but are extensions and expressions of it – not a means to some further end, but are ends in themselves. Consumers persist in sustaining their commitments, often at the expense of their social wants and natural needs, because they deeply care about certain ethical issues – be it animal welfare, human rights, or the environment. By acting upon their deeply cherished concerns, they actualise their unique moral identity and allow it to transpire in the way they conduct their lives, as individual agents and as social actors operating in an inter-relational, intersubjective, and interdependent reality.

Limitations and further research: beyond individual identities

So where to go from here for deepening the understanding of ethical consumers and further exploring the multiple dimensions of their hybrid, complex, and fluid identities? This book has placed its primary emphasis on exploring personal aspects of identity formation, consisting of one's reflexive exploration of the self, its subjective concerns, commitments, and practices. It thus provides an individual level analysis of the ethical consumer identity, with a focus on the personal emergent properties – reflexivity, self-awareness, intentionality – as they manifest themselves in the context of people's engagement in ethical consumption. Whilst exploring the inner psychological process whereby consumers' ethical self is produced is an important and long overdue undertaking, it should not escape our attention that consumption in general and food consumption specifically is rarely a purely individual experience, but more often a social one (Carù & Cova, 2003; Warde, 1997). In a bid to counterbalance emphasis on the subjective practices and personalised choices of individualistic consumers, Cherrier (2007, p. 323) highlights that "an ethical consumption experience goes beyond an individual act in the marketplace … Consuming ethically links consumers to family members, friends, the state, and the market". Indeed, as my research has shown, individuals' performances of ethical consumption, while motivated and guided by their distinct personal identities, are deeply embedded in an expansive complex network of social ties and relationships. This means that the formation of ethical consumer identities does not occur in isolation or some kind of a social vacuum, but involves continuous negotiations between the self and the social world.

That the self is shaped by influences originating from both inside and outside sources is an axiom among many identity theorists. Lawler (2008, p. 5), for instance, strongly negates the view of individuals' uniqueness – their true or deep self – as "something which belongs to the person in question and has nothing to do with the social world". Identities, she insists, should be seen as "formed *between*, rather than *within* persons" (Lawler, 2008, p. 5, emphasis in original), not in opposition, but *by* the social world. These ideas are closely echoed – and amplified – in Archer's account of identity formation which, while placing a strong emphasis on the notion of reflexive self-production, nonetheless recognises that "the inner conversation cannot be portrayed as the fully independent activity of the isolated monad, who only takes cognisance of his external social context in the same way that he consults the weather" (Archer, 2003, p. 117), and that the reflexive reasoning "is shaped by the networks of relations within which it takes place because these profoundly affect what *does* and *can* satisfy the subject and be sustained by each of them" (Archer, 2012, p. 97, emphasis in original). Relating the argument to the domain of ethical consumption, Cherrier (2007, p. 323) argues that "the key reference points for constructing ethical consumption lifestyles come from not only the inside (self-identity) but also the outside (collective identity)". It is on this basis that she advocates approaching the questions of identity formation from a dialectical perspective that is sensitive to the fact that identities are both individually and socially produced: "the notion of identity does not emerge from an individual process of self-identification and therefore, should not be regarded solely as individualistic" (Cherrier, 2007, p. 329). It is, therefore, important to consider that the decision to take up the ethical consumer identity does not arise solely on the basis of self-enquiry, but results from processes of identification and recognition by others. Although none of the ethical consumers who took part in my research have explicitly framed their practices in terms of collective participation, there are, no doubt, many committed individuals for whom the meaning of consuming in an ethical way goes beyond personal choice to give rise to the feelings of shared morality and collective identity. Thus, further research, while continuing to acknowledge the heterogeneity of ethical consumers and the diversity of subjective motivations and meanings surrounding their behaviours, should examine ethical consumption as a communal activity, a shared domain of moral values, principles, and beliefs and explore the collective identities that emerge and develop within it.

Likewise, this book took an individual-level approach to consumer emotionality designed to illuminate the ways in which emotions and concerns link up to propel consuming agents into ethical actions. However, analysis of the role of emotionality in shaping consumer intentions and practices cannot be complete without exploring its collective and social aspects. In this respect, valuable contributions have previously arisen from Gopaldas' (2014) study, which approached consumer emotions from a sociocultural perspective and examined the role of marketplace cultures and ideologies in structuring patterns of consumers' affective response. Echoing Gopaldas, I suggest that future research should focus on bridging the individual and social levels of the analysis of consumer emotions and more closely investigate the relationship between socio-cultural factors – marketplace forces, political ideologies, moral discourses – and consumer-level emotional reactions. Turning the spotlight on shared concerns and emotions as well as social interactions, whereby they are spread and exchanged, can offer a means to broaden our view on the various ways in which people develop into and as ethical consumers.

The scale of debate on ethical-consumer behaviour is extensive and multi-faceted at both the individual and social levels. By producing a sociological account of ethical consumption that takes people's "first-person view of the world seriously, both recognising their agency and what their concerns tell us about them and their situations" (Sayer, 2011, p. 10), I aimed to contribute to this debate and steer it towards a more balanced approach, providing both micro-level and macro-level explanations of ethical consumer practices and identities. My findings highlight that the processes of change for agents and structure unfold in closely interrelated ways: the evolution of ethical consumerism occurs in social contexts which themselves change as a result of the actions and choices of consuming agents. This suggests that analysis of consumption phenomena needs to move both upward towards a more extensive view of social structures and downward towards a more nuanced grasp of the motivations and actions of individual agents. Through examining the morphogenetic cycle (Archer, 2000) underlying the formation and transformation of consuming agents and their social contexts, future research can progress towards a fundamentally historical understanding of ethical consumption. Further, acknowledging that identities are both individually and socially constructed, and understanding the external contexts of people's internal mental and emotional workings will allow developing a dialectical perspective on the formation of consumers' ethical self. It is by following these avenues that future research can provide a more comprehensive, multidimensional, and multi-level view of ethical consumer practices and identities.

References

Adams, M., & Raisborough, J. (2010). Making a difference: ethical consumption and the everyday. *The British Journal of Sociology, 61*(2), 256–274.

Archer, M. (1998). Realism and morphogenesis. In Archer M., Bhaskar R., Collier A., Lawson, T., & Norrie, A. (Eds.), *Critical Realism: Essential Readings* (pp. 356–381). London, UK: Routledge.

Archer, M. (2000). *Being Human: The Problem of Agency*. Cambridge, UK: Cambridge University Press.

Archer, M. (2003). *Structure, Agency, and the Internal Conversation*. Cambridge, UK: Cambridge University Press.

Archer, M. (2012). *The Reflexive Imperative in Late Modernity*. Cambridge, UK: Cambridge University Press.

Barnett, C., Cloke, P., Clarke, N., & Malpass, A. (2010). *Globalizing Responsibility: The Political Rationalities of Ethical Consumption*. Oxford, UK: Wiley-Blackwell.

Bhaskar, R. (2010). *Reclaiming Reality*. London, UK: Taylor & Francis.

Carù, A., & Cova, B. (2003). Revisiting consumption experience: a more humble but complete view of the concept. *Marketing Theory, 3*(2), 267–286.

Cherrier, H. (2007). Ethical consumption practices: co-production of self-expression and social recognition. *Journal of Consumer Behaviour, 6*(5), 321–335.

Giddens, A. (1991). *Modernity and Self-identity*. Stanford, CA: Stanford University Press.

Gopaldas, A. (2014). Marketplace sentiments. *Journal of Consumer Research, 41*(4), 995–1014.

Haidt, J. (2012). *The Righteous Mind*. New York, NY: Pantheon Books.

Lawler, S. (2008). *Identity*. Cambridge, UK: Polity Press.

Sayer, A. (2011). *Why Things Matter to People*. Cambridge, UK: Cambridge University Press.

Warde, A. (1997). *Consumption, Food, and Taste*. London, UK: Sage Publications.

Index

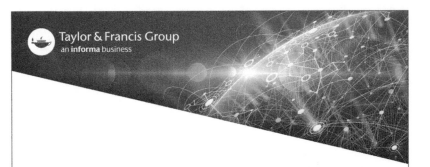

Printed and bound by CPI Group (UK) Ltd, Croydon, CR0 4YY

01/11/2024

01782621-0005